D. GARY YOUNG

The World Leader in Essential Oils

Young Living Essential Oils, LC
3125 Executive Parkway
Lehi, UT 84043
USA
800.371.3515
YoungLiving.com

ISBN 978-0-9863282-7-5

First Edition

Printed in the United States of America

The information contained in this book is for education and entertainment purposes only. Any reference to health, nutrition, diet, and food products should not be used to diagnose, prescribe, or treat any condition of the body and should not be used as a substitute for medical counseling. Neither the author nor the publisher accepts responsibility for such use.

D. GARY YOUNG

The World Leader in Essential Oils

YOUNG LIVING™
ESSENTIAL OILS

Gary teaching members on a
Young Living Trip to Egypt.

Table of Contents

PREFACE

This is a book about the life of D. Gary Young and the path that led him to become the World Leader in Essential Oils.

A lot is written about aromatherapy and the use and application of Mother Nature's precious essential oils. Many books have the word "aromatherapy" in the title, and people all over the world are buying them, looking for alternatives for gardening, cooking, skin care, emotional and physical well-being, environmental concerns, and other interests.

However, with all the information available, little is written about where these oils come from and how they come to be found in soaps, perfumes, cosmetics, food flavorings, and the little glass bottles that people purchase for their personal use.

Who knows where the plants, herbs, trees, etc., are grown; and who knows anything about distilleries and the extraction of the oils? Most importantly, who understands about their quality and the need for absolute purity? Even people who visit a distillery may see the plant material being loaded, the boiler fired, and the oil coming out of the separator. But the intricate details of making sure the distillation process is exactly right for the best extraction is unknown or elusive to most.

This is a book that tells that story and answers those questions that perhaps have never been asked. It's a book about one man, his gifts, and his gift to the world—Seed to Seal, which is the story of the essential oil, from the planting of the seed, to cultivating, harvesting, extracting, testing, distributing, and educating.

This unexpected path began because of an accident that took away from him the only life he had ever known and led him down an unknown path that gave him purpose and a reason to live. That purpose led him into the annals

Gary is thrilled to see the beautiful Juniper oil.

of history and to the far corners of the earth, where he was continually learning more about essential oils and discovered fascinating lost knowledge that he was driven to share with God's children.

His company, Young Living Essential Oils, is the vehicle through which his vision has been taken to the world, wherein millions of people have been blessed in numerous ways. This is the true-life story of one man's dream filled with adventure, heartache, pain, triumph, and success beyond measure. May his passion for life touch your heart and help ignite the fire within you.

Gary teaches distillation to the members attending Silver Club at the farm in Mona, October 2015.

ACKNOWLEDGMENTS

When I started this book, I had no idea how complex it was going to be. Writing about someone's life is one thing, but showing that life of over 60 years in pictures became a huge challenge. Just finding the pictures, let alone with good or acceptable resolution, was often times like looking for a needle in a haystack, especially for the earlier years. I could not have completed this book without the dedication and tenacity of the individuals mentioned here.

John Whetten sat with me hour after hour, day after day, hunting for pictures, placing them, rearranging text and photos, and working Photoshop magic on many of the old pictures. With the use of his excellent language skills, John also helped with the editing and some of the rewriting. He has traveled the world with Gary and often knew where and when, was there when "it" happened, and took many of the pictures that we used. He is extremely talented, dedicated, and willing; and I cannot thank him enough.

David Petty is a very talented graphic artist, who has a wonderful eye for photography. David probably knows the archived pictures better than anyone and spent a lot of time finding "just the right one." His beautiful pictures are seen all through the book, and his clever photo editing enhanced so much of what you see. He has been with Young Living for many years and has been an invaluable part of creating this book.

Alene Frandsen has the final word before anything goes to print. She meticulously read through the manuscript many times, catching all those "little mistakes" and inconsistencies, as well as editing the grammar and sentence structure. As a former business educator, she rewrote awkward sentences and paragraphs and clarified what didn't make sense or could be expressed in a better way. Her editing skills are amazing and saved us all a lot of time and frustration.

Karen Boren, researcher and technical writer, documented dates, places, and explanations. She searched the Internet and many history books to make sure that our information was written correctly and expressed in the proper scientific way. As she edited, she found errors that most people would not see and has helped in many difficult areas with her writing skills.

Paul Springer used his beautiful graphic design skills to create the final layout and look of this book. He has been doing design work for us for many years and understands the nature of Young Living, our mission, and what we wish to convey to the reader. He has been a friend for a long time and has designed many of our publications, including Gary's historical novel, *The One Gift*.

My Family really wondered if I would ever finish. Gary helped so much with specific areas of writing in telling his story. He helped find particular pictures and edited for content, specific dates, and places, as everything had to be accurate going back so many years, as only he could do.

Gary, Jacob, and *Josef* were often frustrated with me as I stayed up late writing and editing, but I know this will be a thrill for Gary, especially to see his legacy in such detail, and for the boys to have this history of their father. I love them and appreciate them very much.

Clint Walker stars in the TV series *Cheyenne* (left), 1955-1962, and in the movie *Yellowstone Kelly* (right), 1959.

Clint Walker at the Young's home, June 2013.

Clint Walker, with his wife Susan, Gary, and daughter Valerie, receives the Spirit of Young Living Award at the International Grand Convention, 2013. While on stage, Clint, in his admiration for Gary, said, "Gary is like a modern-day Moses, leading the people of the world from chemical bondage to the freedom of natural products and the discovery of essential oils."

FOREWORD

My life has been an exciting and colorful adventure but full of tremendous pressures and demands that have taken their toll physically and emotionally. I have spent almost 45 years of my life in the movie industry in grueling days of filming with long hours that demanded tremendous mental focus. It was always exciting to see the film or the series finish, but I often felt exhausted, so I was open to things that offered longevity and increased energy.

While in Hollywood, I was told about a young man who had recovered from a terrible accident that was supposed to have left him in a wheelchair for life. However, this young man had a very strong will and had discovered many natural things and some of Mother Nature's secrets that he said helped him. It was amazing to many that he was able to get out of his wheelchair and, even though it took a long time, was able to walk again.

I, too, was interested in the gifts of Mother Nature, and so I decided to go and meet this young man to see what I could discover. I had grown up helping my father in his health food store, so I had some knowledge of herbs and different plants and was always interested in learning new things.

When I met Gary Young, I was impressed with his tremendous positive attitude. I had no idea that he was always in pain because he never let anyone know. He never talked about the life-threatening accident that destroyed his physical strength and powerful body.

Only when we became friends on a deeper level did he express some of his innermost feelings and how he was constantly looking for things that would help him on his road to recovery.

He shared his excitement about his discovery of essential oils and his feeling about their value to mankind. We talked about things that we both had experienced in the world of health and wellness and our mutual feelings of exploring Mother Nature's world. He was always so encouraging and supported my ideas and desires to try new things.

I watched Gary develop his farms and grow his business and was so honored when he invited me to speak at his convention and tell my story and just be a part of what he was building.

As I read through this book, which reflects over 40 years of our friendship, I am deeply impressed and touched with his dedication and unwavering commitment to his mission. Growing up in poverty with little education, getting out of a wheelchair and learning to walk again, and moving forward with his dream against what the world would say was impossible is surely a tribute to his unbreakable spirit and love of God, which has been his foundation.

We love the essential oils, the Ningxia Red, and so many of the products he has developed. They all fit right in with our lifestyle.

Susan and I have enjoyed our association with his family and have delighted in watching Jacob and Josef grow up, ride horses, and be a great example of what their parents have taught them.

I am grateful that I have lived long enough to see Gary's vision become a reality and that Susan and I have had the blessed opportunity for many years to benefit from what he has created. www.clintwalker.com

— Clint and Susan Walker

With his trusted horse, Sundance, that brought him years of enjoyment, riding in rugged mountain terrain, Gary reminisces in front of the old trapper's cabin, where he sometimes stayed with his father while hunting in the Idaho wilderness.

DEDICATION

Gary's great love is being with God in the mountains on his horse with his family beside him. Were it not for such a terrible, debilitating accident, he would still be in the mountains, the path that destiny has taken him would never have come to be, and the gifts that he has given to the world would never have been given. His instinctive knowing about the essences hidden within the vast web of the life-giving flora that Mother Nature offers to the world and the quest to discover their secrets would never have entered his mind.

The curiosity of humankind has been on-going since man first walked on the earth, and knowledge has been lost and found over and over as time has passed through centuries of discovery. Essential oils have been used since the beginning of time for perfume and cosmetics, for physical and emotional well-being, and for the most fascinating and deeply spiritual rituals that only the soul of each individual can understand. Only the Father of us all, the Almighty, has all the answers; and He has given us the challenge and the opportunity to figure them out. Throughout the ages, there have been those who would delve into this universal knowledge, just scratching the surface, only to see it disappear into history or to the restrictions of those who would squash this knowledge for profit and power.

In our modern world of the 21st century, there is more knowledge than ever before; and there are those of us who fight for our right to choose our desired way. The path that destiny has chosen for D. Gary Young has been a challenge with much difficulty and opposition, but he has never wavered in his determination to share with the world what he has discovered—the gifts of Mother Nature.

This book is dedicated to all who carry the torch of freedom, determined to live an abundant life of health and happiness and have a desire to share what they have discovered with those who are looking for the messenger who would bring them the same opportunity—the people of the world.

Mary and Gary at the Young Living International Grand Convention in Dallas, Texas, August 2015.

INTRODUCTION

A pioneer in his own right, D. Gary Young has spent the last 30 years of his life researching the ancient ways and traditional methods of distilling from the last century. He has sadly watched the true art of distillation slowly being lost to fast extraction with modern equipment and chemicals.

Plant care has gone to chemical fertilizers and pesticides that have weakened plant immunity and reduced the quality of the oil, causing lower levels of chemical constituents and even losing many valuable compounds. Nutrient-depleted soil is unattended, resulting in weaker plants and loss of quality.

Rural facilities without financial means remain crude and inefficient, often producing a lower yield and lesser quality. Some older distilleries in more remote areas of the world fuel fireboxes with old tires, wood, and often garbage to heat cookers made of carbon steel, which are most undesirable for plant extraction of essential oils.

High-tech laboratories both recreate chemical molecules to add for increased volume and/or manipulate the chemical composition for a more "pleasant smell" for a greater marketing advantage.

A true, unadulterated, pure essential oil as God intended is difficult to obtain, as even many growers and distillery operators are lured into the practice of adulteration because of the desire for money and power.

From the time of his accident and his discovery of essential oils, D. Gary Young has been committed to his research and the true art of distillation. Having used the oils to support his own personal journey and that of many others, he knows the difference between the pure and the adulterated, which has made his path very precise. Purity and quality of the oils are most important to him, regardless of the time and money that it takes to produce the desired results.

This book is a compilation of many things happening in many parts of the world, often at the same time. An exact chronology is impossible to write as new activities continually transpire at the same time in different locations. The reader will jump back and forth from farm to farm with the development of the crops and the distillery.

This historical journey also recounts the early years and life experiences of D. Gary Young that have given him the foundational experience and knowledge for the path that he has taken and how he acquired so many skills that have enabled him to accomplish so much in the development of his farms, distillation, and becoming the world leader in essential oils.

The farms have been visited by tens of thousands of people, and many have participated in various aspects of harvesting, distilling, bottling, and seedling planting and reforestation.

Some of the pictures are not the best resolution because they are old and taken with old cameras, and because of the low resolution, they cannot be enlarged; but I thought a lower quality picture was better than no picture.

May you enjoy your educational experience in learning about how aromatic plants are grown and distilled and come to a greater understanding about the decision you make when you choose to use an essential oil.

Mary Young

Mary Young
November 2015

20 years ago, D. Gary Young established the quality system we call Seed to Seal®, which makes Young Living the most unique essential oil network marketing company in the world. Seed to Seal is the process that controls the production of every essential oil from the farm to the labeled bottle ready to be sold. The Seed to Seal values are the foundation on which our Young Living global corporate offices, farms, and cooperative farms operate. Young Living uses a five-step Seed to Seal protocol for our own farms and our partner, cooperative, and vendor farms worldwide.

The Young Living Seed to Seal process ensures the highest quality possible of every essential oil poured into the bottle at our production facility, thus providing our members with the confidence that they are buying the very best essential oils available.

SEED

We guarantee that every essential oil is pure and authentic. We work with many botanists, agronomists, and other scientists to identify the plant genus and species to ensure that the seeds, cuttings, and plants used in cultivation are the correct ones and will produce the greatest essential oil benefits documented by historical use and scientific research.

CULTIVATE

Good cultivation practices are critical for all essential oil-bearing plants that are grown for the extraction of Young Living essential oils. Individuals gathering wildcrafted plants are expected to be eco-responsible for same-species collection. The resulting essential oil will be rejected if it does not pass the rigorous scientific testing in the laboratory.

Young Living ensures the highest quality essential oils by controlling and documenting the health of the soil through proper composting, natural fertilizers, crop rotation, and the amount of water through rain and/or irrigation, as well as considering daily sun and temperatures.

Young Living tracks the country of origin and the species for every essential oil as part of the Seed to Seal process. We periodically visit partner, cooperative, and vendor farms and distilleries to document current practices to ensure the best techniques are used.

DISTILLATION

Young Living distilleries use only stainless steel and glass material for the distillation vats, condensers, and separators. We document all pertinent distillation data, including volume of plant material, curing time if needed, steam temperature, time of distillation, essential oil yield, and harvest location of every batch of essential oil at our farms. We distill only single species of plants for the essential oil extraction and must, therefore, have plant verification before entering into any partner agreements.

We expect documentation from partner farms of steam temperature, distilling time, essential oil yield, and field conditions so that we are able to trace back to the source if needed.

We also support ancient and modern techniques of steam distillation, hydrodistillation that is used for the extraction of the oil from tree resin, and cold pressing of the rind of fruits to produce the highest quality

TEST

Our essential oils must be completely natural and unaltered from the single-species distillation. Young Living does NOT accept any diluted, cut, or adulterated essential oils; and no chemicals are allowed to be used in the distillation process.

Young Living tests incoming essential oils with as many as nine different tests: specific gravity, optical rotation, refractive index, FTIR, and polar and non-polar column Gas Chromatography (GC). If questions arise, then further testing may include HPLC, GC/MS, chiral column GC, ICP-MS, and IRMS (isotope ratio mass spectrometry). This last test can detect nature-identical synthetic compounds. These tests ensure that the oils meet our quality standards.

A Certificate of Analysis is created for every essential oil lot Young Living receives. This document includes all of the analytical data from the previously mentioned tests, the botanical name of the plant, and its

SEAL

After the oils are tested and approved by our Quality Control department, they are ready for bottling. Every time oil is ready to be bottled, it is first batched with a number, tested again, and then sent to the bottling process. If there is ever any question about a particular oil, it is easy to go to the batch number for more specific data, even to being traced back to the distillation from which it came.

After bottling, the oils are ready to be shipped to members all around the world.

Life was harsh, but it made him strong, determined, and fearless.

A PATH UNKNOWN

"The time will come that if you don't grow it, you won't have it."

How could Gary have understood the significance of these words spoken by Mr. Henri Viaud? The Frenchman was speaking about his beloved lavender essential oil; but when he said this, lavender was plentiful.

That was in 1991 when Gary was studying distillation with Mr. Viaud, who at that time was considered the "father of distillation" in Provence, France, the lavender capital of the world. Tucked away on top of his mountain, Henri Viaud, a retired professor of mathematics and physics at a university in Toulouse, France, was known to have distilled everything, even rocks, in his life's work to explore all avenues of distillation for the advancement of understanding the process for the best extraction of essential oils known to man.

What did Mr. Viaud feel or sense about the future of what was so precious to him? Did he know that there would be a scarcity of oils with a demand that could not be met? Did he know that essential oils would become so popular that because the demand could not be met, they would be adulterated and made synthetically to eventually be used in everything from food flavorings to cosmetics and even soaps, a far cry from their amazing and intended benefits?

Sadly, families that had been distilling for centuries are closing down their distilleries. They are being dismantled and the property is being sold as real estate for housing, restaurants, and other business ventures. Some of the land is also being sold for commercial use, while other crops are being taken out and replaced with quicker cash crops such as hemp, soy, canola, etc. The young people are heading to the cities for the "technical life," while pure oils are being endangered by growers and distillers willing to cut corners for profit.

But how did Gary Young, a farm boy from the mountains of Idaho, come to be in France studying distillation with the famous Henri Viaud? Why was he driven to learn everything he could about essential oils, how they were produced, and their benefits to mankind?

The story goes back to perhaps another lifetime in comparison—back to the mountains where Gary grew up, his immigration to Canada, and his dream to forge a life out of the wilderness. He could never have foreseen the change that would come into his life that would take him down a strange and unfamiliar path, a path that would change millions of lives for the better.

His destiny? For D. Gary Young, the world became his destiny—his teacher, his classroom, his encyclopedia, his destination—his life. Yes, his life revolved around discovering how to heal his broken body, riddled with pain, which would lead him into the annals of history and an ancient world shrouded in the mysteries of essential oils that had been lost to the modern world.

Gary is in front of his maternal grandfather. Nancy, Gary's older sister who moved to Canada with him for a short time, is to the left.

The Donald N. Young family, 1963.

At age 14 Gary tracked and shot this cougar, which had been killing their sheep for over two months.

Gary grew up in a 16 x 20-foot cabin with a dirt roof until age 4. Then his father built a 30 x 30-foot, four-room cabin, where he lived with his parents and five siblings until he was 17 years old. They lived 12 miles from town and had no electricity or running water. Electricity finally came when he was 16 years old, as well as their first hand-crank phone that was routed through the forest service line that connected to the lookout and ranger stations for fire patrol.

Life on the ranch, 1955.

During the summer of 1967, Gary logged over 2,500 miles in the saddle in the Idaho wilderness, patrolling for fires, clearing trails, and packing supplies to the lookout posts.

Gary's father in front of the first dirt-roof 16 x 20-foot cabin, where Gary lived until age 4.

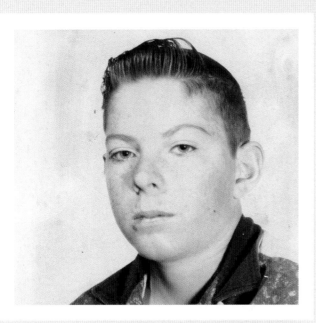

"As a young boy, I dreamed of going to Canada and homesteading a ranch on the Canadian frontier.

"In 1967, when I was 18, my dream came true when I loaded my truck and my horse and moved to British Columbia. In 1968 I got in on the last homestead act in the Caribou District with 320 acres 30 miles in the wilderness, where I began building my horse ranch and logging business.

"Little did I know that here in the wilds of Canada, the destiny that awaited me would change my life forever."

In September 1967, at age 18, Gary packed his red mustang and drove to Canada for the first time. A few weeks later, he returned to get his truck, his horse, and the rest of his belongings, intent on living his dream.

The original 16 x 20-foot cabin that Gary built on the property he homesteaded 30 miles outside of Quesnel, B.C., 1968.

Gary's logging business flourished.

Learning to walk again; a rare photo since Gary refused to have a picture taken of him in a wheelchair or using crutches or a walker.

Destiny Speaks—A Terrible Accident

At 24 years of age, on February 8, 1973, Gary suffered a very severe logging accident when a cut tree sheared off and hit him in the head, resulting in 3 open skull fractures, a ruptured spinal cord in 3 places that was classified as an incomplete break, 11 ruptured discs, 16 broken and/or crushed vertebrae, a broken pelvis, the right scapula broken in 9 pieces, a severed brachial plexus, and 19 broken bones, which included all of his ribs on the right side and several on the left side.

Suffering with intense pain with the use of only his left arm, he was confined to a wheelchair with a medical prognosis that he would never walk again. His life was one of 13 drugs, including morphine and an antidepressant, and a world that seemed dark and hopeless with no light at the end of any tunnel.

After two failed attempts at suicide, he sunk into an even deeper state of depression. He had no insurance and slowly lost everything—his logging equipment, his ranch, and his livestock. His wife took the children and left him, unable to cope with his current condition and the burden of the future.

In his third attempt, he decided to "fast" himself to death; it was the only way out. No one could rescue him or force him to eat; but after 253 days of only drinking water and lemon juice, the most unexpected happened—he felt movement in his right toe. The doctors explained it as one of those unexpected medical phenomena. They suspected that because of his fasting, his body did not get the needed nutrients necessary to manufacture scar tissue, enabling some nerve endings to reroute and reconnect, which was the beginning of many years of intense pain in his struggle to walk again.

He reconciled himself to living in misery and hopelessness and to finding a new path of life where he could survive. Because he couldn't go back to logging and farming, his mind began to expand as he started exploring different avenues of healing and the world of books.

He stopped all his medications to clear his mind and began experimenting with different modalities and many avenues of study. Attending a community college, going to massage school, running his own health food store, and growing in his knowledge of herbs and natural healing, his amazing road to recovery brought attention from many people wanting help.

Gary's emotional recovery was dramatic. Told in his own words, his story gives us a slight glimpse into his life at that time and is a tremendous insight into his anguish, deep pain, and determination.

"The lowest ebb in my life was when I tried to commit suicide and failed, not once, but three times. I had no place to retreat to when I realized I couldn't even die. Never before had I felt so disconnected from God, my family, and the world. I felt so rejected and was sure that God hated me because He wouldn't even let me die. I had nowhere to turn.

"Had it not been for my father, who came to my hospital room, and shaking his finger at me, told me to grow up, quit feeling sorry for myself, accept the hand I had been dealt, and figure out a way to get on with life, I might have continued to wallow in self-pity, depression, and eventually succeed in ending my life. But my father knew my temperament, and he pushed the right buttons, called 'tough love.'

"After speaking those words in a tone of disgust and anger, he turned and walked out and never came to visit me again. He made me so angry as he walked down the hall, never looking back, while I was bawling and crying out to him, 'That's easy for you to say. You don't know what it's like to be paralyzed and in constant pain with no drugs that help. If it's the last thing I do, I will get out of this bed, and I will walk and ride my horse again!'

"When I finally got out of the hospital to go home for a weekend, I discovered that my family was on welfare, which sent me deeper into depression, and I cried, 'No, I will not be on welfare; take it back!' My wife asked me what she should do. The ranch was gone, the horses and cows were gone, the logging equipment was gone, and there was no more money.

"Alone, I cried for hours, and no one came to sympathize with or console me. When there were no more tears to run down my face, I started to think about what I could do. I asked my mother if my paint box and canvases were still in storage and if she could get them for me. Perhaps I could earn a living by painting; at least one arm worked. I started that night and by the end of the next day had painted a beautiful, scenic picture of the mountains.

"Everyone loved it and expressed their feelings of support. I continued to paint and catch rides on weekends down to the strip mall parking lot, where I could sit and sell my paintings. Soon I was getting jobs to paint murals in lawyers' and doctors' offices and homes. But in a small town, there was not enough business to support me as an artist for very long, and so the jobs became fewer and fewer.

"One night while reading the paper, I saw a job listing for a truck driver to haul chips. A friend drove me to the mill office; but when I told the owner what I wanted, he laughed and asked me how I was going to drive a truck in a wheelchair.

Unable to work, Gary sold his paintings on a street corner in town on the weekends to earn money after having lost everything because of the accident.

Gary again began to make a living in the logging industry by driving a truck retrofitted with a hand clutch and brakes.

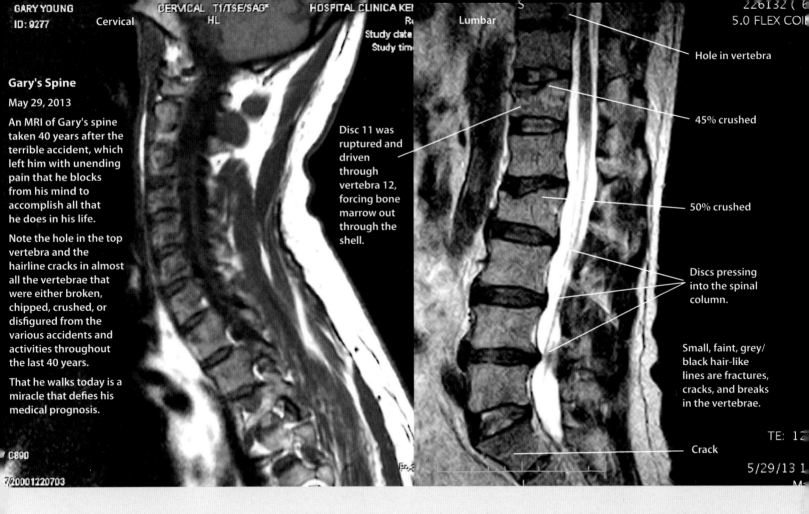

Gary's Spine

May 29, 2013

An MRI of Gary's spine taken 40 years after the terrible accident, which left him with unending pain that he blocks from his mind to accomplish all that he does in his life.

Note the hole in the top vertebra and the hairline cracks in almost all the vertebrae that were either broken, chipped, crushed, or disfigured from the various accidents and activities throughout the last 40 years.

That he walks today is a miracle that defies his medical prognosis.

Disc 11 was ruptured and driven through vertebra 12, forcing bone marrow out through the shell.

Hole in vertebra

45% crushed

50% crushed

Discs pressing into the spinal column.

Small, faint, grey/black hair-like lines are fractures, cracks, and breaks in the vertebrae.

Crack

TE: 12

5/29/13 1

C890

720001220703

I told him that if he equipped the truck with hand controls, I could drive it. With a dubious look, he pointed to a Mack day cab tractor in the yard and told me that if I could drive it over to a semi chip trailer, hook up to it, and drive it back to the office, we would talk about a job.

"I wheeled my chair through the gravel yard to the truck, pulled myself up on the steps, pulled my chair up beside me, and shoved it behind the cab. I got in the seat and started the truck, and once the air pressure was up, I released the brakes, turned the truck off, put it in gear, and started it again.

"I pulled around in front of the trailer, turned the motor off, put it in reverse, and started it again, backing under the trailer. When it locked on the bull pin, I set the brakes and turned off the motor at the same time.

"I slid out onto the step, lowered my chair to the ground, and then slid along the frame so that I could hook up the air hose and power cord. Then I slid back to my chair and wheeled myself to the landing gear handle, cranked it up, wheeled back to the step, and got back in the truck.

"I started the engine, built the air for the trailer, released the trailer brakes, and turned the engine off. Putting it in the lowest gear, I started the engine again, coasted down to the office door, shut it off, set the brakes, and got out over the steps into my chair. It took me almost an hour with climbing up and down and pulling my chair up and down. As I started to wheel myself to the owner's office, he came out to meet me and, with tears in his eyes, told me I had a job.

"On my first trip, when I backed the trailer up the unloading ramp, I was concerned that the foreman would get upset with me for taking so much time to unhook, dump, and rehook my trailer, making the other trucks wait. I moved as fast as I could, sliding out, lowering my chair, and grabbing the rod that I had made so that I could reach under the trailer to hook the fifth wheel release lever and trip it.

"I noticed him watching me and thought that my job might not last. I wheeled back and started to crank the landing gear down when he came out of the control tower and down the steps. To my surprise, he told me to get back in my truck and said he would take care of hooking and unhooking me whenever I came.

"It was amazing how a feeling of wholeness started to come over me now that I was working again."

Gary's office in 1983 in La Mesa, Mexico, Baja California.

This is the second motel that Gary renovated to use for his research facility for $1,300 per month.

RECOVERY
Mexico and a New Path of Discovery

He moved to Southern California, where he continued with his education by enrolling in a Naturopathic college. At the same time, he opened a a small office in Chula Vista, California, and built a research center for physical and emotional well-being in Rosarito Beach in Baja California, Mexico.

His single desire was to help people find answers and solutions to their own problems. Through 13 years of constant debilitating pain and frustration, Gary went from a wheelchair, to a walker, to crutches, to a cane, to very painful, slow walking. But walking again, as painful as it was, kept him determined to discover new possibilities of healing.

During this time, Annemarie brought her sister from Switzerland, who was not doing well, to the research center to see if she could find some help. Annemarie had grown up working in the natural health products industry, had her own laboratory, and formulated various health products. She had been studying essential oils for many years and was

1985

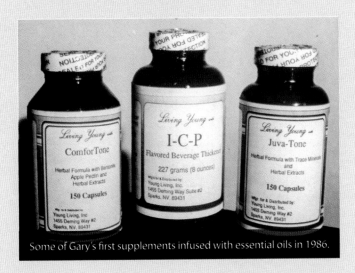

Some of Gary's first supplements infused with essential oils in 1986.

very interested in their health benefits. She gave Gary some research that she had translated from French about essential oils because she felt he would be interested and then invited him to attend a conference that was taking place the following week in Geneva, Switzerland, where medical doctors were presenting their research on essential oils and their effects on respiratory illness.

That was the beginning of his path that has led to thousands of discoveries about essential oils and the immense possibilities they offer for physical, emotional, and spiritual application.

Gary had fun recording Donna Riley, Clint Walker (from the movie series *Cheyenne*), and his wife Gigi singing together while at the research center.

The marquee in Bonita, California, in 2008, still showed Gary's *Young Life Wellness Center*.

Gary's laboratory, where much discovery and learning took place.

Gary's research center was beautiful and there was always time for relaxation and entertainment.

The office manager organized Gary's original apothecary, where many herbs were ground from plants gathered on weekend excursions.

Gary welcomes Dr. Bernard Jensen and his wife Marie, 1986.

Essential Oil Research in France

Gary was invited by the French medical doctors to travel to France and join them as they made rounds in the hospital in which they were conducting their research. Gary was fascinated and driven to learn more. His questions were endless, as this new world continued to open his mind to greater understanding and new possibilities.

He returned home from France with 13 oils and began experimenting with them to discover and learn more about their usage and application. There were no books available written in English about essential oils, let alone anything on usage and application; and the French medical doctors were publishing in journals that were only in French.

This truly put Gary on the frontier of essential oil science as he began to discover what oils and oil combinations to use, what worked better, and how to apply them. With no written information in the U.S., no Internet, and virtually no one with any conclusive experience, the frontier was his. So he boldly went forward, against tremendous criticism and

warnings, determined to unlock the hidden mysteries of this ancient science that he instinctively understood.

He even began infusing his herbal formulas with essential oils and was impressed with the increased efficacy of the supplements. He felt so much excitement that he wanted to share his findings with the world.

But the world was not ready for his discoveries; and people laughed at him, slandered him, and told him it was impossible to do what he was doing. But the few who had been using Gary's new products infused with essential oils began to spread the word, and more and more people wanted what he had.

He decided to sell his research facility and start his own marketing company. The demands were increasing and he felt a new passion growing inside. He could teach many more people at the same time than he could by helping one person at a time, day after day. The experience he had gained was tremendous, and now he wanted to teach and share his new knowledge, to begin a new chapter.

Dr. Jensen, well-known for his research in health and natural healing, presents Gary with one of his most popular books.

Gary was fascinated with live cell and dark field microscopy.

He did a lot of experimenting with the first oils he brought home from Europe and made many exciting discoveries. His thirst for more knowledge heightened with a desire to study distillation. With his farming background, he had a natural interest in the growing, harvesting, and extraction of the oils from aromatic plants, adding a new path of research and discovery. It seemed that France was the place to go, but he didn't know anyone there and didn't have the slightest idea how to make a contact, besides the fact that the French were not open to sharing their "secrets." French lavender was their claim to fame, and they didn't want anyone else, especially an American, getting involved or interfering.

Gary flew to France a couple of times a year, determined to learn about distillation. He carried a backpack and a sleeping bag, prepared for anything. After all, he grew up in the mountains, and comfort was not a consideration. In a small rental car, he drove the countryside looking at the fields and distilleries from the roadside. He found a few people who spoke English, but there was little they would tell him, which created a lot of frustration.

Typical old French distillery, no longer in use.

There was a lot of excitement as Young Living started to grow.

Young Living headquarters in Spokane, Washington, 1989.

Gary was a guest speaker at many expos and seminars.

THE BIRTH OF YOUNG LIVING

When Gary started his new business, he was a "one-man band." From Monday through Thursday, he answered the phone and took orders and then packed them and had them ready for shipping. He filled the oil bottles by hand in his little lab and hand wrote the labels for them. He did everything he could; and that which he couldn't do, he contracted out piece by piece.

Thursday evenings, he would put the answering machine on and fly out to a different city such as Phoenix, Los Angeles, or Dallas and then travel around with his members and friends like Eldon and Nancy, Loretta, and Anna-Maya, doing meetings to grow his new little business.

Sunday night he returned home to get ready for all the orders that he hoped would start coming in on Monday.

Once a month he would spread all the orders out on the living room floor and calculate the commissions by hand. That continued for months until the business had grown enough that he felt he could start hiring a few employees.

It was truly a simple beginning with real intent to give something to the world that he loved and that he believed would benefit mankind.

In August 1992, even with a broken ankle, Gary took the first group of members to France to see the lavender fields and distillation.

Gary with the Landel family: Jane, Jean-Noël, Véréna, and Nicolas, in the backpack, 1991.

Nicolas now works with his father and Young Living on the farm in France; 24 years later, Nicolas teaches visitors about distillation, 2014.

Jean-Noël Landel—The French Connection

In 1990 Gary was lecturing at a Whole Life Expo in Anaheim, California, and had a group of members running the booth. Jean-Noël Landel was visiting from France, hoping to find someone to whom he could sell his lavender oil. Since 1985 Gary had been experimenting with oils and infusing them into his supplement formulations and had some of them with a few single oils on the display table. As Jean-Noël walked by the booth, he saw them and stopped to talk. He was still talking when Gary came off the stage. They quickly discovered a mutual interest, and Jean-Noël invited Gary to come to France for a visit, not thinking he would really come.

One month later, Jean-Noël was absolutely shocked when Gary called him from Paris. Gary didn't know when he would travel, so he didn't contact Jean-Noël ahead of time to let him know he was coming. Jean-Noël and his wife, Jane, who is an American, welcomed Gary into their home, which was

the beginning of a great and lasting friendship that opened so many doors for Gary in the world of the French lavender growers. Their daughter, Véréna, was four years old; and their son, Nicolas, was just six months old. Jean-Noël and Jane had never used the oils in their home for their personal needs; and so when Nicolas was crying and spitting up and would not go to sleep, they didn't know what to do. Gary held Nicolas in his lap and showed Jane how to massage his feet with lavender oil until he fell asleep. Gary also taught Jane how to do Vita Flex to enhance the use of the oils.

Today, in 2015, Nicolas is a young man, working with his father, Gary, and Young Living on various projects for Young Living and is learning everything he can. Little did Jean-Noël and Jane know that they would embark on a life-long friendship with Gary that would involve many exciting adventures of discovery.

The first Young Living cooperative French distillery in Simiane-la-Rotonde has now been upgraded with stainless steel and more efficient technology.

Gary, Dave (Bodil's husband), and Jean-Noël load the cooker with lavender at the farm in France.

This distillery was in full operation when Gary started studying distillation in France in 1991.

Gary received some of his first distillery training with this portable French distiller, which he purchased in 2006 and is now on display at the farm in Mona, Utah.

Philippe Mailhebiau—New Spiritual Insight

On one of Gary's trips to France in 1991, Jean-Noël introduced him to Philippe Mailhebiau, who was writing and teaching about essential oils. Philippe had a laboratory and was conducting research on both the physical and emotional aspects of essential oils.

Philippe opened Gary's mind to the emotional aspects of the oils, which Gary found fascinating, since he believed the oils carried a spiritual essence that greatly influenced how an individual responded to the oils. Studying the emotional and spiritual characteristics of the oils was soul-stirring to Gary, opening another world of learning.

However, Philippe was not a farmer and had no distilleries, so he could not teach Gary anything about growing and distilling, but Philippe and Gary shared a mutual interest in discovering the unknown benefits of the oils. They became good friends and Gary invited Philippe and his associates to come to Idaho to teach the members, which was a rewarding time for everyone.

Philippe visited the farm in St. Maries.

Philippe opened Gary's mind to the emotional aspects of the oils, 1991.

Marcel Espieu—Mentor and Friend

Gary traveled back and forth many times learning about France and the essential oil industry from Jean-Noël. One year after Gary's first visit, Jean-Noël introduced him to Marcel Espieu, the president of the Lavender Growers Association, who was not too open or receptive to this curious American who kept asking questions. However, with time, Marcel could see how serious Gary was and eventually invited him to see his distillery. Gary volunteered to work filling the firebox and doing anything that Marcel asked. Gary stayed late into the night keeping the firebox filled, and Marcel would then come to trade places with him until morning.

Marcel began to trust Gary more and more and developed a friendship that was destined to last a lifetime. Marcel taught Gary everything he could, and Gary would often drive with him to take the lavender oil to the perfumeries in Grasse, France. It was amazing that even though they did not speak each other's language, they were able to communicate through working together and accomplishing the desired goal.

Gary is sitting in front of Marcel at the French distillery.

As Gary learned more about distillation, he began to visualize building his own distillery. Everybody laughed and Marcel made jokes about the American who thought he could do what the French have done for hundreds of years. But for Gary, once the dream was there, it was as though it were already done.

Gary, Mr. Viaud, and Jean-Noël talk about distillation.

> Trust in the
> LORD with all
> your heart and
> lean not on your
> own understanding;
> in all your ways
> acknowledge Him,
> and He will
> make your
> paths straight.
>
> –Proverbs 3:5-6

...llation

...ther teacher of ...wanted to meet. ...that time was considered the "father of distillation," lived in the mountains in Provence, not far from Jean-Noël; but he was not open to having visitors, especially "curious" Americans. However, Gary persisted; and one night, five years later, Mr. Viaud came to Jean-Noël's house for dinner when Gary was there.

In a very short but intense interview, Mr. Viaud asked Gary what essential oils meant to him. Feeling tremendous pressure and with much anxiety, Gary said, "I believe that essential oils are the closest physical and tangible substance that carries the spirit of God on earth." Pointing his finger at Gary and in a dramatic tone of voice with a heavy French accent, he said, "You are right and anyone who messes with them should be treated like a criminal." Then he stood up, turned, and abruptly walked out the door.

Jean-Noël walked Mr. Viaud to his car; and when Jean-Noël returned, he said to Gary, who was feeling very defeated, "Mr. Viaud wants you on his mountain at 6 a.m." It was an exhilarating moment for Gary that left him sleepless and with great anticipation.

Gary spent many days with Mr. Viaud, cutting and harvesting wild lavender, thyme, and rosemary and then distilling them. He had only two 1,500-liter cookers, so the distilling was slow. The days added up to weeks and then months as Gary traveled back and forth from the U.S., going up the mountain to Mr. Viaud's distillery to work. Between Mr. Viaud and Marcel, Gary could not have had better teachers.

Jean-Noël, Mr. Viaud, and Gary; serious camaraderie.

Gary traveled all over France with Jean-Noël meeting growers and visiting distilleries. Jean-Noël and Gary leased their first farm in 1992. Then in 1993, they found an abandoned government farm that had a lease/purchase offering that seemed to be perfect for them. But after two years when the time came to buy, there was just too much paperwork, and the government was too difficult to work with; so Jean-Noël started looking elsewhere.

Over the length of four years, Gary traveled back and forth to France, spending as much time as he could studying and working in all aspects of the production from planting, to harvesting, and to all facets of the extraction process. Volunteering to help harvest, working in the distillery, and tending the firebox and boiler at night enabled him to study the cooking and distillation in detail. He made meticulous notes as he measured and drew plans. He learned everything he could with a "knowing" feeling inside that one day he would build his own distillery.

Jean-Noël and Gary at the second farm that was leased from the government in 1994.

INSPIRATION FROM AROUND THE WORLD

The Ancient Secrets of Egypt

The more Gary learned, the more he was driven to go to Egypt, the heart of the ancient world of essential oils. His experiences were amazing and the knowledge he brought home convinced him even more that there was so much that had been lost to the modern world.

Gary traveled throughout Egypt exploring the pyramids, tombs, temples, museums, and visiting government offices. He even wandered through the marketplace looking at all the essential oils, dried herbs, resin burners, beautiful oil containers, etc., to get ideas. The information was voluminous. He felt very comfortable "walking back in time" as history told its story, opening up still another chapter of discovery.

In 1991, during Gary's first trip to Egypt, he was fortunate to meet Dr. Radwan Farag, dean of the biochemistry department at the University of Cairo, who had been conducting research on essential oils for years. As of 2015, he has published nearly 200 research papers. Dr. Farag willingly shared his knowledge with Gary. He was a wealth of information and a wonderful mentor.

Gary tells an interesting story when he visited the Temple of Isis on the Island of Philae:

"One of the guards came to me and said, 'I know what you come here to see' and indicated that I should follow him. It was very strange and I thought the guard just wanted money, so I joined a tour group to get away from him. But the guard was persistent and tapped me on the shoulder a second time and repeated himself. I tried to get away from him by going a different direction, but he came around the corner and met me face to face and for the third time said, 'Come quick; follow me. I show you what you come to see.'

Dr. Radwan Farag visited Gary's lab in Riverton, Utah, 1995.

"By then I was annoyed but had also become a bit curious, so I followed him. He had an old metal key to a heavy iron gate that he quickly unlocked, pulled open, and motioned for me to hurry inside. As soon as I went through the gate and started up the stairs, the guard walked in and locked the gate behind him.

"It was very unnerving, but I didn't feel I should go back; so I followed him up the stairs into an open area on the roof and over to a room with no door. As the guard motioned for me to enter, I could see the sunlight illuminating the hieroglyphics on the wall showing the ancient ceremony of the 'cleansing of the flesh and the blood' using essential oils. It was a glorious moment of discovery for me that became the foundation for the 'emotional clearing' that I have used in my teaching."

Dr. Farag was amazed at Gary's experience and translated the idea of the ceremony for him. Dr. Farag was fascinated with Gary's interest in essential oils and in his development of growing aromatic plants for distillation. In 1995 Dr. Farag came to Utah as a guest speaker and to see the tiny Young Living operation. He was helpful and encouraging and has remained a friend since that time.

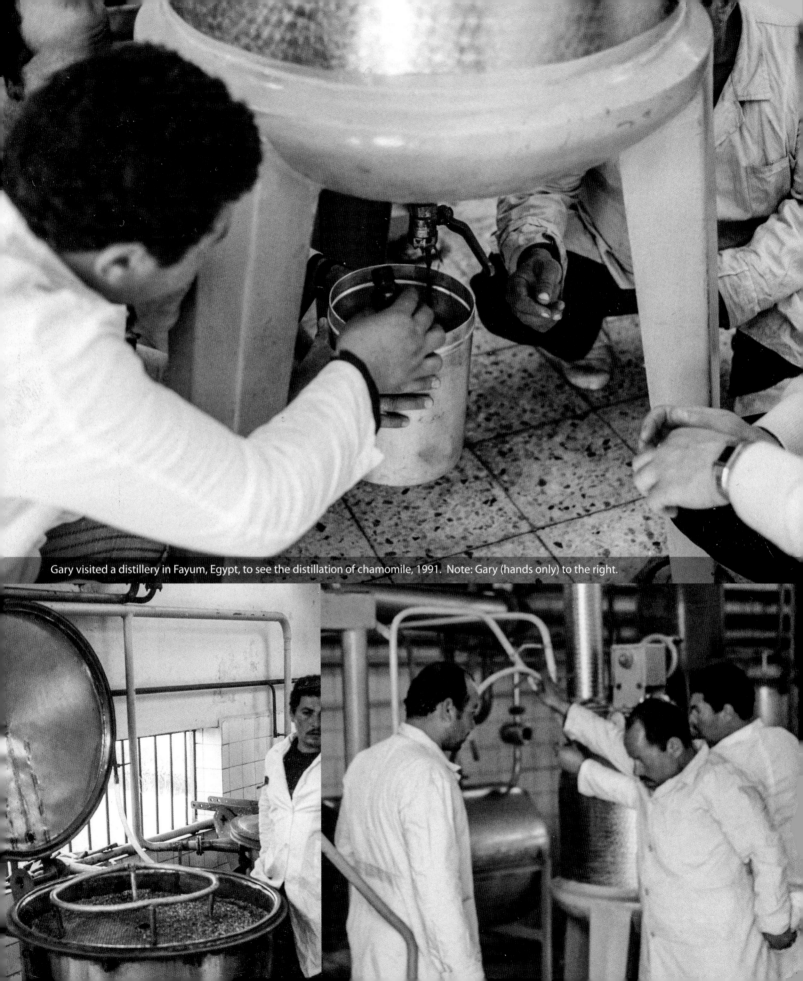
Gary visited a distillery in Fayum, Egypt, to see the distillation of chamomile, 1991. Note: Gary (hands only) to the right.

A guard tapped him on the shoulder and said, "Come quick, follow me. I show you what you come here to see."

In 1991 at the Temple of Isis on the Island of Philae, Gary stands at the doorway of the room where he learned about the long-lost Egyptian ritual known as the cleansing of the flesh and the blood.

Two Egyptian priests stand at the head and foot of the person undergoing the three-day cleansing ritual that included the use of essential oils. This hieroglyph is found in the Temple of Isis in a room rarely shown to the public.

This is an ancient clay essential oil extractor that dates back to 350 B.C., now displayed in the museum in Taxila, Pakistan. The guard in the museum told Gary there were no distillers, but there was an old type of "water purifier" made of clay in a glass case at the back of the building.

Discovery in Pakistan

When Gary and Mary went to Oman in 1995, Gary decided to go to Hunzaland to interview the old people and to see if he could discover their secrets to longevity. Gary had heard that there was a museum in Taxila, a small town outside of Lahore, Pakistan, that had some ancient clay distillers. They had to fly to Lahore on the way to Hunzaland, so it was not much out of the way. Gary really wanted to see these early clay distillers, so that was even more of a reason to go there. It was an exciting moment when he spotted them at the very back of the museum. The guard allowed Gary to take pictures so that he could share his discovery with everyone at home.

The Old World

The old world of Egypt, ancient Arabia, Israel, etc., is filled with remnants of distilleries, essential oils, and apothecaries where salves, beauty creams, and medicines were made. It was part of their culture, and the oils were prized commodities used only for royalty and those of a higher social status.

Essential oils were procured in many different ways as was evident by the different types and methods of extraction that could be determined from the ruins and what was left of the distilleries. It was extremely interesting and educational for Gary, as he was constantly following legends, rumors, books, maps, and stories that he heard from the people in the countries he visited.

Gary talks with both the old and young people in Hunzaland.

In 1990 Gary first traveled to Masada, located on a high mountain plateau on the west bank of the Dead Sea. There he found the ruins of an ancient distillery that was very distinguishable in the rock. As he analyzed the process, it was amazing how his explanation brought it to life. All of this certainly went into his data bank for his plans to build a distillery.

Gary brought back the first lavender seeds from France in 1985 and planted them three years later behind the Spokane office on a quarter acre of ground.

Gary's second-year lavender (*Lavandula angustifolia*) was harvested and distilled in the kitchen.

THE MODERN-DAY FATHER OF DISTILLATION

In 1988 Gary sold his research center in Baja California and moved to Reno, Nevada, to start his new business. A year later, at the end of 1989, Gary moved his headquarters to Spokane, Washington, where he devoted himself to his research and growing his Young Living business. On a quarter acre behind his building, he planted the small amount of lavender seed that he had brought home from France.

The seeds germinated and grew into beautiful, healthy plants, exuding the most intoxicating aroma. As the plants matured and grew bigger each year, Gary was anxious to see if he could produce oil from them.

Gary's dream of distillation started with two pressure cookers welded together for his first distilling experiment in 1991.

As life took Gary down a different path and eventually into the essential oil world, he ventured into many unknown areas, which became his world of discovery. He had taken many notes while helping Marcel, and the memory was vivid in his head. As the ideas of distillation occupied his mind, he began drawing them on paper and mentally fabricating how he would build a distiller.

In 1991 for his first distillation experiment, Gary welded two pressure cookers together, cutting holes in the bottom of the top container to allow water to be poured in and steam to rise up through the holes into the top cooker. He then cut his beautiful plants that were only two years old and by French standard not quite mature enough to produce good oil. However, Gary felt he could not wait another year to determine if he could produce lavender oil.

Distillation on the Kitchen Stove

With the two pressure cookers welded together sitting on the stove top in the kitchen, Gary poured the water into the bottom through the holes that he had cut and then packed the plants tightly into the top cooker. The gas stove was lit and as soon as the water began to boil, the steam started traveling up through the plant material. Shortly thereafter, the first drop of lavender appeared, and Gary's new chapter of distillation was written.

These small cuttings produced 3 milliliters of the most exquisite lavender oil, which put more ideas into Gary's head. A good friend, Dr. Kurt Schnaubelt, analyzed this first lavender oil; and even Kurt was surprised at the quality of the oil from these two-year-old plants. It was Kurt's enthusiastic response that motivated Gary to keep going forward with his dream.

Young Living members were willing and eager to help, 1991.

Gary built two more experimental distillers, each one a little bigger than the previous one. The second distiller built in Spokane in 1992 had a 25-liter cooking chamber, enabling him to distill a larger amount of lavender.

He experimented by distilling pineapple sage (*Salvia elegans*) and common sage (*Salvia officinalis*), which he had also grown in the garden plot behind the office. The pineapple sage smelled nothing like pineapple as he was hoping, but the fragrance of the sage was beautiful. As his curiosity increased, he decided to plant thyme.

All of the crops that were planted in the garden in Spokane and grown from seed, including the lavender seed from France, were the original starts that Gary eventually transplanted to the farm in St. Maries.

The second distiller was built in 1992 and was more advanced in design.

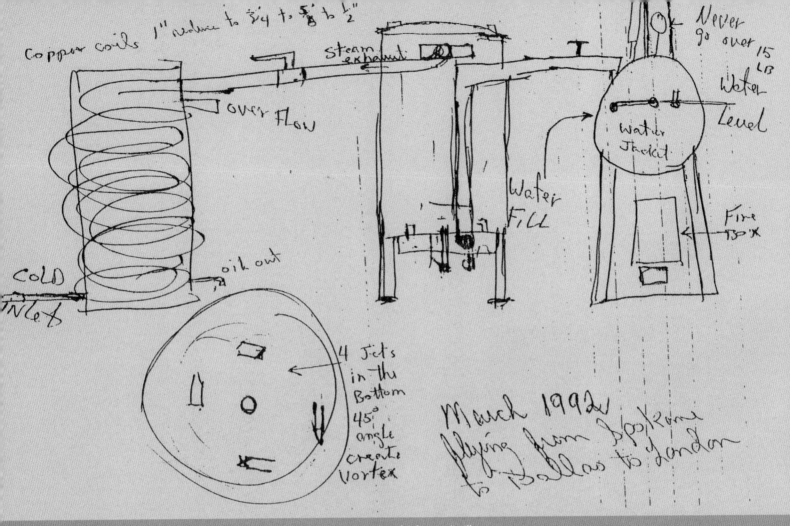

Copper coils 1" reduce to 3/4 to 5/8 to 1/2"

steam exhaust

over Flow

COLD INLET

oil out

Water Fill

Water Jacket

Never go over 15 LB

Water Level

Fire Box

4 Jets in the Bottom 45° angle create Vortex

March 1992 flying from Spokane to Dallas to London

After Gary's first kitchen/stove-top distiller, he began sketching new ideas for future distillers.

Simple Distillation

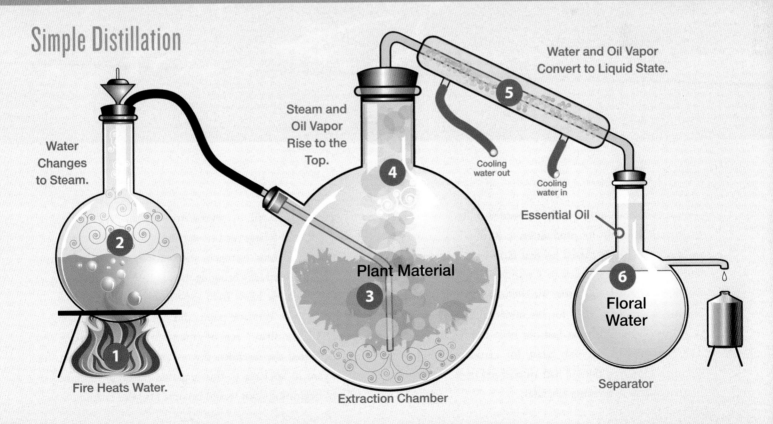

Water Changes to Steam.

2

1

Fire Heats Water.

Steam and Oil Vapor Rise to the Top.

4

Plant Material

3

Extraction Chamber

Water and Oil Vapor Convert to Liquid State.

5

Cooling water out

Cooling water in

Essential Oil

6

Floral Water

Separator

In 1992 Gary began plowing the ground at the St. Maries farm, which had not been worked in over 50 years and had never had chemicals put on it. The soil was especially hard, with a pH of 4.5. It took 10 days to plow 40 acres with a 250 hp tractor and an 8-bottom plow. Burning the weeds, breaking the sod with the Dyna-Drive, disking, harrowing, and leveling the soil were done twice during the summer and twice during the spring of 1993 before planting.

St. Maries—Farming Begins

It was obvious to Gary that he needed more land in order to grow more plants for production to further his research, so in May 1992 he purchased his first farm of 160 acres nestled among the trees in Benewah County in the mountains of St. Maries, Idaho. The farm was remote and untouched by chemicals. It was perfect for his needs, and he aggressively began to till the land that had not been plowed in 50 years. The land had grown only grass for cattle pasture in the summer. However, the soil had an acid pH of 4.5, which was not conducive to growing lavender.

He had to till in several tons of manure, microbes, enzymes, and foliage feed to start bringing up the pH. Every year more calcium, nitrogen, and phosphate was plowed into the ground, gradually bringing the pH to 7.5, the desired level for lavender. He knew from experience that he had to build the soil, since aromatic plants did not grow well in conifer-type soil. He tilled 1 acre of ground in front of the barn and transplanted the lavender, thyme, and peppermint from his backyard in Spokane to this "gigantic" farm, as it seemed to Gary, beginning what would become his great farming legacy.

Gary injected liquid enzymes and microbes that are critical to building the soil for sustaining good crop growth. This was the first field of clary sage.

Gary spent 1993 preparing the farmland and increasing the soil pH, so he could plant small crops of peppermint and clary sage that he planned to harvest and distill. He designed a firebox boiler with a 250-liter distilling chamber and had it built in St. Maries in the machine shop at Fleet Parts and Service. Gary went several times to the shop to give direction and check on the fabrication. He was very exacting and wanted to make sure that it would work the way he wanted.

It was a 30-minute drive of great anticipation when Gary hauled it out to the farm. He quickly hooked up everything to prepare for distillation. Peppermint was the first plant distilled, and it was exquisite.

Bob Skinner, the owner of the machine shop, said, "It was fascinating to listen to them discuss how they were going to build this distiller, but I was more astounded when I knew how much peppermint was distilled and all the excitement that Gary felt when just a few drops of oil fell from the spout into the quart jar that was sitting on a tree stump. It was obviously a great triumph for him."

The distiller worked so well that Gary's vision began to grow bigger as he saw in his mind more crops planted, bigger extraction chambers, and more liters of oil. He was soon drawing the plans to make his vision a reality, and the farm expanded with that reality.

Members volunteered their time to help plant the clary sage and lavender starts that came from the garden plot in Spokane; memorial weekend—38°F. Without the members, it took 12 farm hands 12 hours to plant 1,800 starts.

The water was pumped with a small 5 hp fire pump running 7 Rain Bird sprinkler heads that the irrigation company said wouldn't work, 1993.

Gary's third portable, wood-fired distiller, built in St. Maries, was a good prop for the ol' country doctor in the Steven Segal movie *The Patriot,* 1998.

The first clary sage field and Gary's PVC irrigation system that he built in 1993.

Gary dug out the pond on the farm for a reservoir to catch the rain and the snowmelt runoff and pumped the water with a 5 hp Briggs and Stratton fire pump to the fields through 2 inch PVC pipe with risers and Rain Birds.

Chuck, the owner of Dickerson Pump and Irrigation Co. in Spokane, said it wasn't possible; but after Gary called him to tell him that it was working, he drove all the way to the farm to see what he thought was impossible. Chuck became a lifetime friend and helped Gary with the irrigation needs of the farm throughout the years.

For years, that little pump served the farm very well, but as more crops were planted and the farm expanded, the pump finally had to be replaced with a much larger pump and wheel lines to meet the increased irrigation demands.

Cutting peppermint by hand, separating out weeds, and carrying it to the truck was "back-breaking" work. Clary sage is being irrigated in the background.

Members helped move PVC sprinkler lines.

When Jack and Carma Young managed the St. Maries farm in 1994, Aunt Carma was often out in the field working with everyone else.

The First Harvest

The first harvest was very exciting as the lavender was packed into the 3,500-liter distiller. It was a tremendous accomplishment and everyone was thrilled with the 4 liters of oil produced. The farm was growing as Gary's vision expanded, and it became obvious that more cookers were needed. Different crops were being distilled, and fields were increasing. Soon three more cookers were added; and by 2006 a total of eight cookers were in operation at St. Maries.

The Farming Learning Curve

Clary Sage

Through trial and error, Gary learned a lot about aromatic crops. Clary sage grew extremely well, but the deer loved it and ate it down to the ground. One summer evening, Gary and Mary flew to Spokane; and by the time they arrived in St. Maries, the sun was gone and it was dark. As they drove up the hill and the car lights hit the field illuminating across the plants, more than 100 pairs of eyes stared into the lights, obviously disturbed by these "intruders" who had interrupted their eating delight. Clary sage was like an aphrodisiac to them.

Gary used electric fences, air cannons, dogs on leashes, different scents, every means possible; but there was no way to keep the deer out of the fields. It was hopeless, so Gary decided to grow clary sage only in Utah. Interestingly, those well-fed does that next year had the highest rate of twin births ever recorded by a fish and game office.

Thyme

Most of the thyme plants looked alive but didn't seem to be growing. Further examination revealed a very disappointing discovery: the thyme plants were just sitting on top of the dirt. They had no roots because the gophers had been feasting. Perhaps they loved how it made them feel. Who knows, but that was also hopeless.

Peppermint

Peppermint grew well, but when it rained, the grass grew faster. Organic practices were becoming more challenging; and with the cool nights and early frost, the menthol in the peppermint did not reach the level Gary desired; so he decided that peppermint would be grown only in Utah.

Gary was thrilled with the first clary sage crop, 1993.

Collecting clary sage seeds by hand, 1993-1995.

Gophers enjoyed eating the roots of the first-year thyme, 1997.

Gary cut the peppermint with a small sickle-bar mower—like a giant lawn mower. After the cutting, starts were dug up to replant in a larger field.

The beautiful first peppermint crop went moldy because of the rain and too little sunshine—another learning experience. The crops grew extremely well; but fighting deer, rain, and mold made it impossible to grow some of the crops that Gary had hoped would be oil producing, 1994.

The hot climate in Mona was perfect for the peppermint, and it grew very well. The production was tremendous and the aroma was exhilarating. To our great relief, the clary sage also thrived and the deer didn't have any interest in eating it. The plants grew with their vibrant fuchsia color, the aroma was exquisite, and the production was wonderful.

Gary designed a conveyor system for the peppermint and spearmint so that as the plants were harvested and cut with the swather, they dropped onto the conveyor that carried them up and into the truck that was driven alongside the harvester, and when full, driven to the distillery. It worked very well and greatly increased the efficiency of the harvest.

In 1994 Gary began to prepare the ground for the new distillery site.

The First "Real" Distillery—1994

In the summer of 1994, Gary began the construction of his fourth distiller, but this time it was to be a "real" distillery. He logged and peeled the white fir trees growing on the farm, prepared the ground, rented equipment, pounded the logs into the ground with the bucket on the excavator, and began construction of his first stationary stainless steel, vertical steam distiller.

He had only one 3,500-liter extraction chamber in which the first lavender, clary sage, peppermint, thyme, and tansy were distilled. Because it took so long to steam clean after each different crop, Gary could see the need for more cookers. The distillery grew as Gary had more cookers fabricated to meet the demand of the increasing amounts and different types of plant material that needed to be distilled.

Gary cut and sized the logs for the support beams of the new distillery.

This land was later transformed into a lavender field and then eventually planted in melissa.

Gary worked in the garage that was the perfect place for fabrication, and it was only two minutes away from the site.

Gary's early design for steam injection that he continually redesigned and is very different today.

Gary built with logs because the concrete was too expensive.

Again, a one-man band, but now, almost 21 years later, hundreds of people work in the distilleries around the world.

The wet, distilled clary sage weighed about 1,200 pounds when lifted out by the tractor, but that definitely was not going to work in the future.

As Gary opened the valves and the first steam sizzled through the pipes, the cries of triumph from Gary and Uncle Jack were heard around the farm.

Seeing the first steam was a moment of great triumph for Gary.

The old 25 hp boiler that Gary bought for $500.

Gary found an abandoned 25 hp boiler that was left to rot in a farmer's field outside of St. Maries, which he bought for $500. He had it cleaned and re-tubed and then began assembling everything for the big moment of production. When the boiler fired and the steam sizzled through the pipes into the cooker, the cheering of triumph was heard all around the farm.

Gary was continually thinking about distilling and mentally making new designs for the extraction chambers, condensers, and separators. He had a different idea about the steam delivery into the chamber, so he drew out his plan and had the new delivery design fabricated. He wanted the steam to travel through the plant material in a circular motion so that there would be a more even and better penetration of the steam that would carry the the oil vapor with the steam up and into the condenser.

The water discharge pipes from the condenser were built with PVC pipe because PVC is lighter to move and cheaper than steel; but what a learning experience it was when the next morning, after the first distillation, Gary found that the pipes had collapsed and stuck together because of the heat from the steam and condenser water. Naturally, the PVC pipe had to be replaced with steel.

Every aspect of the harvest and distillation was important to him because he knew that it all had to work in harmony to achieve the best results. Each crop was different so he made many adjustments and modified his system according to what he saw. The early years of harvesting and distilling brought about many changes from the traditional old ways to new, innovative, and more efficient production methods. The more discoveries he made, the more his head spun with ideas, and the more Gary's "art of distillation" evolved.

Mary thought that the aesthetic look of the distillery was wanting, so on went the paint. Even Gary loved the beautification, although it didn't help produce any more oil.

Machinery for harvesting aromatic plants did not exist in the U.S., so Gary was constantly buying equipment that he had to modify. He imported an old lavender harvester from France for St. Maries that worked quite well, making the harvest a dream in comparison to cutting by hand the first year and then by modified sickle-bar mower the second year.

The harvester worked very well for the small acreage in St. Maries; but when lavender went into production in Utah, the need for another machine became necessary. Gary drew the blueprints, and the men in the fabrication shop in Mona went to work to build an even better harvester.

Gary was always designing equipment to make harvest more efficient, and that kept the Utah fabrication shop very busy. He could look at a piece of equipment, see how it worked, and make changes that improved production, often fabricating out of necessity. After all, there was no money to buy farm equipment that cost tens of thousands of dollars; besides, there was little, if any, equipment even available in the U.S. for planting or harvesting aromatic plants.

The first lavender harvester that Gary designed was from a hay sickle-bar mower, which didn't work as he had hoped.

The lavender harvester that Gary imported from France in 2004.

Gary thought that tying the lavender in bundles would make the harvest more efficient; but unfortunately, it proved to be just a waste of time and didn't make a difference. Mary is wearing the sun hat.

Gary and Uncle Jack built a three-row planter in 1994. It worked perfectly and saved a great deal of time and is still in use today, 2015. Note: Mary is sitting on the right.

"Mechanizing"—The Planting Machine

In the beginning, many members came to help. In 1993-94 members planted 45,000 lavender plants in the unusual, freezing cold spring month of May. However, this just wasn't fast enough and would never meet Gary's production goals, so he started designing again. Gary and his uncle, Jack Young, who died in 2001, built the first planting machine that was pulled by a tractor. There were three seats on the back with small platforms next to the seat for the trays so that three people could sit on the machine and plant three rows at the same time, greatly increasing the number of starts planted every day.

It was very rewarding for Gary to go from his crew of eight people, who had worked so hard to plant 1,500 starts by hand in 10 hours, to just three people planting 1,500 starts per hour and one more person operating the tractor.

Tansy

One summer afternoon in 1994 while walking from the house down the dirt road to the distillery, Gary noticed tall, wiry plants with little, yellow, button-type flowers growing along the ditch bank. The aroma was strong and enticing, so Gary decided to distill it and see what it would do. It had a low yield but it appeared that flying insects didn't like it, so it seemed like it would be good for experimenting. Wild tansy was abundant in the fields and ditches, and farmers were very happy to have it cut and taken away, so it soon became a good wildcrafting crop. The animals, especially the horses, loved it when the bot flies came charging.

As Gary watched everyone cutting by hand and using sickle-bar mowers, he realized it was time to look for mechanized equipment and bigger trucks to haul the material.

Tansy was first cut by hand, 1994.

Upgrading to a tractor was a great improvement.

Mary would pack a lunch in the morning, and they'd leave at 5 a.m. for the fields and come back at dark to help with the distilling. Gary operated the chopper, blowing chopped tansy into the truck, and laughed a lot as Mary enthusiastically swathed up and down the hills, even though her rows weren't as straight as he would have liked.

Gary discovered that chopping the tansy didn't hurt the oil and increased the production five times. It was faster, more efficient, and was

The mechanics of distillation are about the same. First, you have the boiler or heating source for the water to produce the steam that flows into the "cooker" or distilling chamber in which the plant material is packed. The steam travels up through the material, causing the plant membranes to open and release the oil vapor, which is carried up with the steam into the condenser, where the steam and oil begin to separate as they convert back from a gas phase into a liquid phase and flow into the separator. Oil droplets bubble up through the separator to the top of the water, so as the oil accumulates, it can be poured off into a container.

The "art of distillation" is in knowing all the details. But how would one even begin without the aptitude and a good teacher? It takes years of experience, and finding an experienced teacher for this "dying art" is a big challenge. A good operator has to know how much water, the water temperature, how much plant material, how tightly it is packed, in what size of an extraction chamber, how long to steam the material, and when to ramp the temperature based on the sound of the steam coming from the chamber, which is a very fascinating aspect of the distillation.

The steam "sings" with a different harmonic pitch as it moves up through the plant material and must be even and constant. If the pitch is not consistent and the temperature is not ramped soon enough or is ramped too fast, the chamber will flood, causing reflux and homogenization; and the oil will fall back into the plant material and usually cannot be recovered. Any oil that is recovered will most likely be scorched and will certainly be missing many important oil compounds.

Every aspect of distillation can make a difference in the oil quality: the design of the distillery, the temperature of the water going into the boiler, and the water purity. Then, of course, the seed quality, the soil pH and nutrients, planting and cultivating, time of harvest, Brix testing, curing time

1. **The water is preheated and the minerals are removed.**

2. **The boiler turns the water into steam.**

3. **Plant material is loaded into the cooker.**

4. **Steam travels up through the plant material, releasing the oil vapor.**

5. **The steam carries the oil vapor into the condenser, where they both return to a liquid state and separate.**

6. **In the separator the oil rises to the top, where it is drained off.**

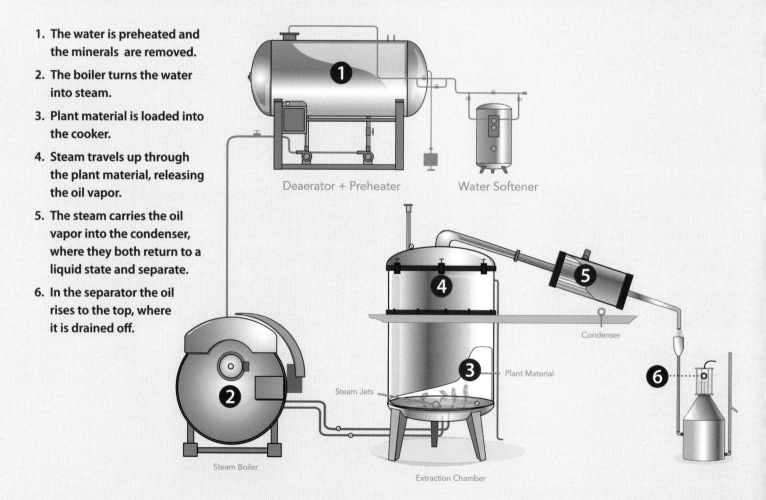

Deaerator + Preheater

Water Softener

Condenser

Plant Material

Steam Jets

Steam Boiler

Extraction Chamber

Convention of 1995 brought members to the farm to see their first distillery produce oil.

if needed, and so many more intricate details all affect the quantity and quality of the oil produced. Distillation is truly an art acquired through meticulous study, experimentation, and knowledge of plants and farming.

Can an individual learn to do this from a book? Perhaps the very basics can be learned, and possibly a distillation could be started; but one will never learn the "art" from a book. Learning the art of distillation takes years of experience with logical thinking, trial and error, learning the many facets of farming, the exactness of mechanical functionality, and an analytical and creative mind.

It is a combination of skills that few people have. Gary learned many of the fundamental skills growing up farming, having to figure out how to fix old equipment and trucks, trying to figure out the best way to get the job done, what worked and didn't work, and what "just made plain, good sense." The fact that he had lived in the mountains and logged with horses was just an added benefit.

There weren't any calculators, adding machines, computers, or Internet. Everything was done by hand or figured in the head with mental calculations, along with hours of hard physical work. All that he learned in his youth gave him a tremendous advantage, as those skills came together because of his passion to learn the "art of distillation."

With every crop, every harvest, and every distillation, more was learned; and the new information increased Gary's knowledge about the many facets of the Seed to Seal process. What wasn't known became very exact as the years of farming and distillation continued. Today, there is less trial and error and more knowing about why different things happen and what to do to solve the difficulties of any given situation.

Perhaps the challenge and opportunity to make new discoveries with a completely new list of questions to be asked and answered is part of what drives Gary's pioneering spirit. His quest to bring new and exciting additions to Young Living is a driving force in his life.

The foundation logs were covered with concrete and steel, 1997.

The Research Farm—So Many Questions

St. Maries was like a research farm. Gary used all that he had learned in France and all that he knew about soil from growing up and farming with his father, which added much to his research; but the plants, soil, temperature, and climate were different for different plants. The French always distilled lavender the same way, the way their families had distilled for years. They had it down to a science; research and experimentation were not necessary—they thought.

But Gary was always looking and wondering, what if? How? Why? He was constantly asking himself questions. What about a different steam injection system? Would this make a difference? Yes, his vortex spiral steam injection system did make a difference. What size cooker was best for what amount of plant material? What were the different needs of different plant materials? Was the distillation yield better when the material was whole, coarsely chopped, or finely chopped? What was the yield with branches and leaves together or separated? How long should the plants be left on the deck after cutting to cure? Should all plants be cured and for how long or distilled immediately after cutting? What was the needed temperature for different plants? How long should the plants be steamed and at what temperature?

The questions were ongoing; and even today with the tremendous accomplishments at the farms and all that has been learned about the plants and the operation, questions continue to be asked. How can we make the planting, the harvesting, and the distillation more efficient? Is there a better way? That is the nature of Gary Young, which is why he continues to make new discoveries that benefit all those in the essential oil world.

The first condensers were vertical and the swan neck was a rigid tube connecting the cooker to the condenser at a 90° angle. The swan neck, now made of flex tube, made it faster to connect and disconnect with simple cam locks, rather than having to bolt and unbolt it.

In Gary's second generation of condensers, he changed the angle to 45°, so there would be a smoother flow to capture all the oil droplets.

More cookers were built as more crops were planted. After the swan neck was changed to a flex tube, it increased the amount of oil captured. Oftentimes, droplets were caught in sharp corners and any uneven edge, so the objective was to make the flow as smooth as possible.

When the plant acreage increased, the need for more distillation capacity increased as well, besides the fact that it took so much time to clean the cookers after each distillation. The cookers were first steamed for an hour, and then soap was put into them and steamed for another four hours. Then they were rinsed and steamed again for another two hours. The water was then drained off, and the cookers were steamed again for another two hours. It made more sense to put in more cookers and designate different cookers for specific plants than to spend so much time cleaning the cookers.

In 2011 the first spa was installed near the separators so that it could be filled with the floral water containing micromolecules from the distillation. This water in the past was just drained off and wasted. Gary kept wondering what could be done to find value for this water. The idea came to him to install a hot tub and fill it with the floral water so that people could relax in it.

Since that time, everyone who sits in this floral water is thrilled with what they experience. Some say it gives them an ethereal feeling, others say it is rejuvenating and energizing, and still others feel a total renewing of their well-being. Adults and children alike are thrilled to soak up those micronutrients.

It is exciting every time the oil bubbles up into the separator.

Every cooker has to be steam cleaned, taking several hours before different plant material can be distilled.

The original St. Maries hot tub was inviting to Jeffrey, but usually people took off their shoes and hats before getting in the water.

Greenhouse Research

In order to expand the fields, lavender starts were trucked from the St. Maries farm until greenhouses were built in Mona, Utah. Although the St. Maries greenhouse was quite small, a lot of research was conducted that provided valuable information about germination, foliage feed and nutrients, transplanting, and the needs of the new plants.

In 1996 Gary began a research project in the St. Maries greenhouse with enzymes and soil nourishment. It was a fascinating project from which much information was gathered for treating the soil and nourishing the plants.

Gary discovered that he could test the glucose (sugar) levels of the plant material with a Brix instrument to see when the glucose was the highest, enabling him to determine the best time to cut the crop.

Higher Brix levels mean that the plants will produce greater volume and the highest quality of oil. Now Young Living farms all over the world use Gary's system for determining the oil potential of the plants prior to harvesting.

Gary moved the two small greenhouses in Spokane to St. Maries in 1992, where he enlarged them, so he could grow more starts and conduct research.

Untreated

Treated

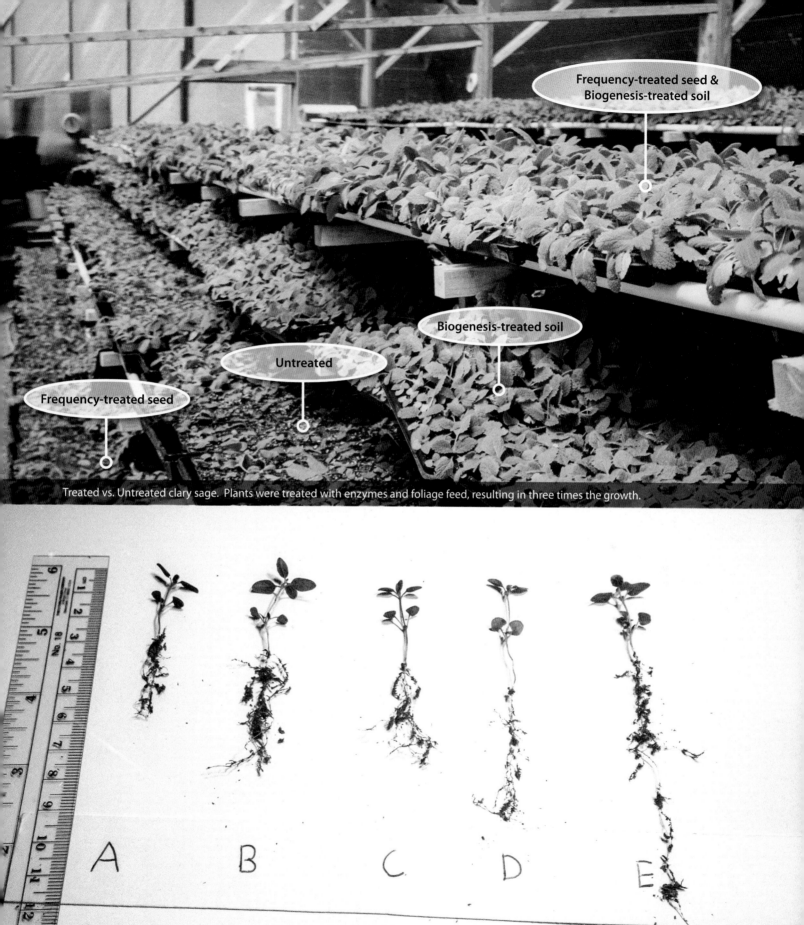

Frequency-treated seed & Biogenesis-treated soil

Biogenesis-treated soil

Untreated

Frequency-treated seed

Treated vs. Untreated clary sage. Plants were treated with enzymes and foliage feed, resulting in three times the growth.

A   B   C   D   E

Plant and root growth were measured and recorded.

The first little garden in St. Maries was in front of the area where the lodge was eventually built.

The first lodge, built in 1999, felt roomy; but with more members coming to the harvest, it began to feel cramped.

The lodge continued to expand and now has a large kitchen, dining and training room, showers, and laundry facilities.

The new addition, built in 2010, seats about 350 people comfortably and is a wonderful place to teach and learn.

The cabins at St. Maries are toasty warm during the winter.

Beautiful first-year lavender grows into big, bushy plants.

The St. Maries shop and distillery at sunset.

Jacob loved St. Maries lavender, 2002.

From left to right: the office and mechanic shop, the distillery, the red barn, the lodge with green roof, the farmhouse, greenhouse with white roof, and small shop behind the greenhouse. Lavender and melissa fields are growing well, and a newly plowed field is ready for planting.

Gary wondered if melissa would even grow in cold, wet St. Maries. The original seed came from France in 1991 but was first planted in 1997 and harvested in 1998. The St. Maries climate was perfect and similar to France; melissa grew beautifully and produced an exquisite oil.

Over 120 acres of organically grown melissa are planted on the St. Maries farm, making it the largest producer of Melissa oil in the world. Tucked away in the mountains of northern Idaho, it is a beautiful place to visit when the melissa and lavender are in bloom.

Members helped keep melissa on the tarp until they carried it to the truck to be hauled to the distillery.

Melissa

Melissa oil is not in abundance in the world, making it very expensive. However, Gary was enthralled with this oil, so he decided to see if it would grow in cool, wet Idaho. It was amazing. Everything was perfect and the plants grew very well. However, the harvest was difficult and didn't produce as much oil as Gary expected, so again he started analyzing and experimenting to discover melissa's distillation secrets.

During two years of harvesting and constantly testing the Brix levels, he discovered the right time for harvesting and curing. Melissa is harvested with a swather that uses a sickle

Members love operating the equipment.

bar to cut the stems of the crops so that they fall neatly into windrows, where they are picked up and transported. But Gary made an unexpected discovery, and that was that melissa did not produce well when cut and left on the ground, so he put tarps underneath the swather to catch the plants as they were cut. Then as the tarps were full, they were lifted onto a flatbed trailer and driven to the distillery.

It worked well and seemed to solve the distilling problem, except that it was very slow and labor intensive. Many members came to the harvest and worked together, walking behind the harvester, loading the tarp, and when filled, pulling it out and getting a new one ready to go. It was a challenge trying to keep up with the large amount of heavy plant material and picking up loose pieces that didn't make it onto the tarp.

It took a couple of weeks to cut the melissa fields, which was too long, pushing past the best time for distilling. Gary knew he had to do something different, so again he began analyzing and designing equipment to mechanize the harvest. He drew his design and began the fabrication in the shop at the farm. He and several others worked on it during the summer but didn't finish it in time for the harvest, besides the fact that Gary broke his hand during fabrication.

Gary designed the conveyor to work with the harvester so that as melissa was cut, it would go directly into the truck, rather than being left on the ground. The harvest went from needing 18 people to 3 and reduced the harvest time from 14 days to 4.

However, the following summer it was ready and went into the field; and within a week's time, the harvest was finished. It went so fast and the oil production was so successful that more acres of melissa were planted.

St. Maries became a fabulous home for melissa, and today the St. Maries farm has become the largest producer of organically grown Melissa oil in the world, along with its exquisite Lavender and wildcrafted Tansy essential oils.

Getting Melissa out of the tarp wasn't easy. Cherié (right) always comes to winter and summer harvests and brings many with her to be a part of Seed to Seal.

Early in 1996 it became evident to Gary that he would need to be able to do testing within Young Living. Third party testing was important, but Gary wanted to know for himself; so he began to look for teachers and places he could go to learn about analytical instrumentation for testing essential oils. He wanted to be able to do the testing himself and be able to read the reports. That began a whole new path of education.

In 1992 Jean-Noël first introduced Gary to Dr. Hervé Casabianca, a highly trained analytical chemist for essential oils in France. Dr. Casabianca is well-known in the industry and helped write the French AFNOR standard for essential oils. He became a mentor for Gary and taught him about how to do the testing and how to interpret the analysis.

This opened Gary's eyes to yet another facet of essential oils. As his interest heightened, he began looking for opportunities to learn more about the chemical structure of oils that would give him the scientific foundation to determine their benefits and enhancements through blending different oils.

Gary completed 120 hours of GC/MS instruction with Dr. Baser.

In 1996 Gary went to the Anadolu University in Eskisehir Turkey, to study with Prof. Dr. Hans Baser, who was teaching a course on gas chromatography (GC) analysis. Gary completed the 120 hours of intense study and was excited about the world of essential oil analysis.

With this new knowledge, he could differentiate between different species of the same oil to understand why one species had greater benefits than the others or why different chemotypes of the same species each had their individual benefits. He found that combining oils could be very specific depending on their chemical compounds, and the response could be amplified.

Blending oils was an aromatic adventure; and with experience, he learned that depending on the compounds of different single oils, he could greatly change or enhance the aroma of the blend. Gary's instinctive understanding of the oils and his fabulous sense of smell gave him an amazing edge in the aromatic world of essential oils.

When Gary returned home, he bought his first GC instrument and hired a microbiologist for his new laboratory that filled a very small room in the old Payson building. Dr. Casabianca came to set up and calibrate our new testing instrument to the exact column-wall thickness used in his CNRS labs in France. This tiny beginning has evolved into a very large laboratory today with millions of dollars of highly specialized instruments for scientific analysis and research of essential oils, as well as scientific equipment for testing other Young Living products.

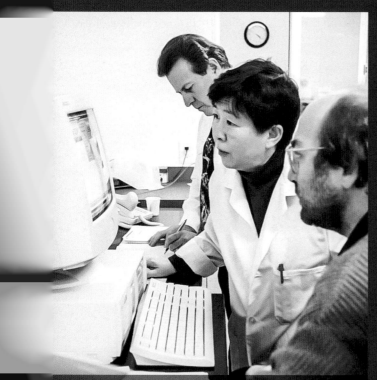
Dr. Casabianca came to Utah to instruct Gary and Sue, biochemist, on the new GC in the Payson office in 1996.

Gary and Chris receive instruction for their continuing education on GC/MS with Dr. Casabianca in France, 2010.

As Young Living grew and more distilleries were built, Gary wanted to have the testing capability on-site to give him immediate information for making decisions about the distilling.

There was a small room in the office in St. Maries that Gary used for testing during the few weeks of distilling in the summer months. But when he started hauling chips from Highland Flats to distill during the winter, more testing was needed.

So in 2012 a whole new room was built onto the St. Maries distillery. A new laboratory with a GC instrument came to the farm so that the oils could be tested for their chemical composition immediately after distilling. This way, Gary had instant information, enabling him to see if he was achieving the results he wanted.

With this on-site testing, Gary is able to determine if he needs to change the distilling time, temperature, or volume of plant material. Some plant material needs to be distilled in smaller amounts; and with the analysis, that can quickly be determined. Both a bottling and a labeling machine were also installed so that members and volunteers could see and be part of the entire Seed to Seal process.

An entire new spa was built with three hot tubs, and five immense holding tanks were installed in the basement underneath to hold the floral water so that it could be pumped into the spa tubs on demand. After working hard all day in the field, relaxing those aching muscles in the soothing floral water that calms and clears the mind is an amazing experience. Just relaxing in the spa and enjoying visiting with other members is one of the highlights at the farm.

Today, in addition to the laboratory in the Spanish Fork Warehouse, Young Living has laboratories at the farms in St. Maries, Idaho; Highland Flats, Idaho; Guayaquil, Ecuador; Fort Nelson, British Columbia, Canada; and Split, Croatia.

The third Young Living farm in Simiane-la-Rotonde, France.

The castle overlooks the Young Living Farm, outlined above.

Some of the first European members who came to help.

YOUNG LIVING SIMIANE-LA-ROTONDE FARM

In 1996 Jean-Noël found a third farm in the Simiane Valley about an hour north of Marseille. It already had lavender growing on it that had not been treated with chemicals; so plans were made, papers signed, and the first foreign-owned lavender farm was bought by Young Living in Simiane-la-Rotonde, France. The farm has 55 hectares (136 acres) and sits in the Simiane Valley looking up at the amazing 12th century castle called Chateau des Agoult, which brings thousands of tourists every year to see its magnificence in the heart of the lavender capital of the world.

Sadly, the new farm had not been well maintained and needed a lot of care; so Gary put out the word, and Young Living members came from around the world throughout the year to help with the planting, harvesting, and distilling. Today, hundreds of members travel great distances to be a part of the Seed to Seal process.

At first, the farmers were skeptical about this American who had moved into their valley. Foreigners were not very welcome, but Gary began getting acquainted; and as they saw his dedication working on the farm and the care that he gave to the plants, they became curious, wanting to know about Gary and his company. As time went on, they wanted to know what Gary thought about farming and why he was adamant about not using chemicals. "Feeding and nourishing the soil will produce better quality oils," was always Gary's motto.

Gary made the statement that one day he would bring seed back to France from the lavender plants growing on the St. Maries, Idaho, farm that had grown from the original seed he had carried from France and planted in 1989 in Spokane, Washington, and in 1992-94 in St. Maries, Idaho. Everyone laughed at such a ridiculous idea. What could an Idaho alfalfa farmer offer a French lavender farmer? Only time would tell.

Members come from all over the world to be a part of the Seed to Seal process.

Gary was saddened by the dying lavender fields in Provence in 1992.

Trouble in France

The farms continued to expand and produce more oil. More and more Young Living members came to be part of the Seed to Seal process. All the farms were producing well except the farm in France. Gary saw trouble on the horizon in the heart of Provence, and now that trouble was staring the French lavender farmers in the face.

From Gary Young, the words were alarming:

"Twenty years ago I saw things changing in the lavender fields in France with the true species *Lavandula angustifolia*. I watched it closely and it wasn't long afterwards when the lavender started dying off.

"When I went back to France in 1985, '87, '88, '89, and '91, when I started working with Jean-Noël, I kept taking seed back home and planting it. From 1992-94, I began expanding my fields at the St. Maries Farm and had enough lavender seed to plant 45 acres.

"Every time I went to France, I watched the farmers try to find out why the lavender was dying. Part of it was an unknown virus that came through the region. Some of the fields developed a fungus, which I believe had something to do with the weakened immunity of the plants after years of growing with fertilizers and being sprayed with pesticides. The lavender just wasn't resistant to some of the bugs and fungus any longer. Then, a severe drought crippled the lavender for about seven years, followed by a one-year reprieve; and then another drought that was even longer took its final toll.

"I had a feeling that one day St. Maries lavender would reestablish the lavender in France and create a rebirthing of true lavender. As we talked about it, there were some laughs; but five years ago in 2010, I brought seed from the St. Maries farm back to France. First, we grew starts; and then two years later, we transplanted them to the field. What was amazing is that this lavender, originally from France and was then returned to France after 20 years in the U.S., was the *only* lavender on our farm in Provence that went into full bloom.

"Our St. Maries lavender was growing in the same field with original French lavender that hadn't died, and there was no comparison. The multiple stems, heavy with lush flowers of beautiful multiple colors from white to deep purple, seemed to dance in the sunlight as they reached farther and farther toward the blue sky. There is not a more beautiful, true lavender than what is growing on our farms now.

"Benoît stated, 'The St. Maries lavender looks like the true French lavender we grew 50 years ago and is the only lavender that produced two crops this year.'

"The other part that thrills me is that this lavender will produce seed that Jean-Noël and Jean-Marie Blanc (production manager) will gather for propagating new starts for next year, starts with strong immunity that will produce the rich Lavender oil that the French farmers remember from years ago. I feel honored and humbled that I can give back to France for all that France has given to me for all these years."

The contrast between the lavender grown from the St. Maries seed (left) that was germinated in a greenhouse in Simiane and the French lavender (right) is most remarkable. Healthy seed and nutrient-balanced soil produce healthy plants, and unhealthy seed and soil produce just the opposite.

Gary operating the first Young Living harvester used on the Young Living farm in France.

Members watch the immense amount of distilled lavender lifted out of the 20,000-liter chamber after distillation.

In 2011 D. Gary Young, with the help of Jean-Noël Landel, his partner of 20 years; Benoît Cassan, the president of the French Lavender Growers Association; and Jean-Marie Blanc, the farm manager who oversees the farm production of planting, cultivating, harvesting, and distilling, merged their farms together as one Young Living Farm to bridge two continents and three decades in the beginning of a bold undertaking: bringing a thriving, healthy *Lavandula angustifolia*—true lavender—from St. Maries, Idaho, back to its origin at the Young Living Lavender Farm in the Simiane Valley of southern France, which today is the largest true lavender farm in the world!

The words of Jean-Noël Landel tell the story so concisely:

"What is really amazing is that this is the only time I've seen true lavender with such a life force in it. Nobody can believe it. The true lavender that has grown from the seed that Gary has brought with him from St. Maries hasn't grown here for over 15 years—it disappeared. But the way it looks now, there is a new future for the lavender farmers in France."

The pipe is laid for the first irrigation system in Simiane Valley.

Jean-Marie, Benoît, Jean-Noël, and Gary merged their farms together to create the largest true lavender farm in the world, 2011.

In 2015, for the first time in the history of Provence, our lavender field was irrigated.

In August 2015, the equipment with the latest technology for harvesting arrived at the farm. When the lavender is cut, it lays for 24 hours in the field. It is then loaded into the new stainless steel vat or distilling chamber and is hauled with a truck to the distillery, where the lid is lowered and sealed. The steam hose is then connected at the bottom, and the distillation begins.

The lid is lowered and the steam pipe is hooked to the bottom of the chamber to begin distillation.

Nicolas and Benoît watch the oil come up in the new separator.

Gary arrived in Riverton, Utah, on Thanksgiving Day 1993. He and a couple of friends drove everything he owned in one pickup truck, one small U-Haul truck, and a trailer that carried all the inventory he owned and a small amount of office equipment. It was a very meager beginning for the new company, Young Living Essential Oils. The dilapidated old building was first leased and later purchased. There was very little working capital and no capital investment from outsiders. Gary eagerly faced the new challenge, solving problems, making it work, and then making it work better.

Gary built the lab and mixed the blends. Rex and LaRue poured the oils by hand, and sticky labels were wrapped around the bottles.

THE MOVE TO UTAH

In November of 1993, Gary moved his company, Young Living Essential Oils, to Riverton, Utah, into an 8,000-square-foot building. Gary built his lab and production room, offices, shipping, and customer service. It was a very tiny beginning with not a lot of promise from the onset, but there was great enthusiasm and determination—and so Young Living began to grow and was incorporated in 1994. Naturally, Gary began to look for another farm; and in 1995 he bought 160 acres in Mona, Utah, where he built greenhouses and planted crops. Uncle Jack, a retired carpenter who lived in Idaho, eagerly came to Utah to help Gary build the greenhouses when Gary called for his help.

Mary in her first "executive" responsibility.

Of the 8,000 sq. feet, only about 5,000 were usable. The building had been used as a mechanic and paint shop, movie house, church, and some "unidentifiable" businesses. The building was leased from a walk-through in the dark; and when the electricity was turned on the next morning, it was a shock to see the mess and deterioration. But that's what it was and so the work began.

Gary did the construction, texturing, and mixing of the oil blends of Abundance and Purification into the paint, while Mary and her mother, LaRue, did the cleaning and organizing.

The first day in business, the orders totaled about $950, but the orders kept coming; and by the end of the first month, the volume had increased to $1,800 a day, but that wasn't enough to pay the bills. So one night after the office was closed, the oil blend of Abundance was sprinkled throughout the office and over the keyboards. To everyone's amazement, the next morning the phones seemed to be ringing off the hook. The excitement mounted and the future looked a little brighter.

This is all there was. There were no big racks, warehouse, or cold storage.

Alene came in August of 1994 and took charge of the office and interviewed and hired new employees.

Alene and Mary put the first product catalog together and started looking at creating new literature.

GOLDEN TOUCH 1   GOLDEN TOUCH 2   7TH HEAVENS   FEELINGS   ESSENTIAL 7

A new "sophisticated" bottling machine improved production immensely, and the overhead bags with packing peanuts really helped get orders out the door faster. A small UPS truck stopped once a day to pick up between 30 and 40 orders.

Within a year's time, Riverton had become too small, and Gary began looking for a new home. Real estate in Salt Lake City was much too expensive, so the old Payson Elementary School became the next headquarters. It was a 45-minute drive south and seemed like a long way, but it was closer to the farm and was a lot bigger with 40,000 square feet, although probably only 30,000 feet were usable. So everything was packed up in the spring of 1995, and this little, growing business moved to the new Payson headquarters.

The building was old and would have been torn down if Gary had not bought it. The wiring for the telephones and computers had to run on the outside of the walls because the concrete walls were just too thick. Many windows were broken and a new roof had to be put on because there were so many leaks. The old kitchen became the lab, and in the first week, a main waterline broke, costing $10,000 in repairs, which was just astronomical at the time. The cafeteria was retrofitted for production, and the gym became the shipping department.

A loading dock had to be built for the UPS trucks.

Rex was excited about new blending equipment that came, and the walk-in cooler was perfect for the oils. LaRue loved the new lab with all the stainless steel. An automated bottling machine arrived from Germany, which dramatically improved production, and one UPS truck was stationed all day at the dock for loading orders. Order Entry was growing, the business was flourishing, and the demand for more oils was increasing.

Order Entry/Customer Service took over the library, and air conditioners were installed in the breakroom for the computers and equipment in IT.

The Utah Farm and Distillery

Lavender, clary sage, thyme, melissa, and many other plants were first grown in the small St. Maries greenhouse from where the first lavender starts were trucked to Mona. When the greenhouses were built in Mona in the spring of 1995, they produced thousands of starts that were not only planted in Utah but were also trucked to St. Maries for the expansion. In 1999 over three million lavender starts were grown in the Utah greenhouses and then transplanted to the fields.

In 1995 Gary purchased 160 acres in Mona and started preparing the land, and Uncle Jack came to help build the first greenhouses.

New starts were transplanted in the spring of 1996.

As more land became ready for planting, more greenhouses were built. The seeds propagated very well, and wheel lines and pipes were bought at auctions and private sales and trucked to the farm so that Gary could begin getting the irrigation system ready.

Gary leased a 1,600-acre farm close to the first farm, which was much more demanding, with more than 800 acres alone of undisturbed sagebrush. The Black Angus Cattle farm had been in the owner's family for years and had been just that—a cattle farm. No chemical had ever been put on the ground, so Gary was excited about this new "organic" farm. The manure was 4 to 5 feet deep in the pens, so Gary bought his first piece of new farm equipment, a manure spreader, and began to spread the manure and prepare the land.

The ground had to be tilled, seeded, and irrigated. He first planted lavender, clary sage, and peppermint; and they all did very well. Gary fabricated a new planting machine three times the size of the first one in St. Maries. He designed and built a new harvesting machine and several other pieces of equipment to make the harvest more efficient.

Gary and Mary went to many farm auctions and private sales, where they found tractors, cultivating equipment, trailers, and all kinds of farm implements. Gary bought his first semi-truck for $3,500, which was used for many years on the farms.

Gary designed and fabricated a six-row planter that was critical for being able to get all the plants into the ground when they were ready to prevent them from becoming root-bound from overgrowth in the small greenhouse cups.

The planter worked very well and hundreds of thousands of new starts went in the ground.

Gary hauled the starts with his semi-truck from the St. Maries greenhouse to Mona until the Mona greenhouses produced thousands of starts that were trucked to St. Maries to meet the expansion there.

Gary and Mary went to many farm auctions, where they found tractors, cultivating equipment, trailers, and all kinds of farm implements. Gary bought an ugly, green, 1973 Peterbilt semi-truck for $3,500, his first since 1978. The truck made many trips after being painted with the Young Living logo and was proudly driven by Gary, who was happy to be behind the wheel of a semi again.

The first automated air seeder for the farm put millions of seeds into small cups for germination in the greenhouses.

The farm had numerous springs that supplied the farm with water.

Over 8 miles of underground pipe were laid for the irrigation in 1996.

Convention in 1995 was a big disappointment for Gary because he had only one cooker to show everyone. But it was obvious that big plans were in the making.

The "straw" is lifted from the cooker, with more plant material waiting to be distilled.

The distillery began to grow when the first cooker and boiler were installed for the first distillation in 1996.

Steam cleaning the first cooker, 1996.

Gary built a small 4,500-liter extraction chamber to distill the first harvest. In 1997 he added larger chambers of 6,500 liters and every year continued to build more chambers to meet the needs of his expanding crops.

The last two 12,000-liter chambers were built in 1999, totaling 12 extraction chambers in which 10 to 15 crops have been distilled over the years, depending on the wildcrafting and what was planted.

Spearmint, German chamomile, hyssop, goldenrod, and thyme were subsequently planted. Each year different crops are planted, and others are taken out, depending on how well they grow and produce oil. Thousands of acres of juniper trees grow in abundance on the neighboring foothills and have been wildcrafted, producing a lot of oil since 1999.

Gary continued to develop both farms as he commuted back and forth between St. Maries and Mona; but when this third farm was finally purchased in 1997 (the second in Utah), the farming of aromatic crops took on a new dimension: more acreage to prepare, more seedlings to plant, more equipment to buy, more inventions and fabrications, more extraction chambers, more employees—just more of everything.

One summer evening, while standing on the porch of the old farm house that was being renovated as the Visitor Center, Gary looked out over hundreds of acres yet to be cultivated and, with that visionary look in his eye, said to Mary, "One day, this farm will be a worldwide destination for people all over the world, welcoming thousands of visitors wanting to learn about essential oils and the amazing Young Living Seed to Seal story."

It was a huge undertaking and money wasn't readily available, but Gary knew it was all possible and went forward with that knowing for what the future would bring.

Peppermint waiting to be distilled, 1997.

The chickens, pigs, and mice that lived in the old farm house had to be relocated, so renovation could take place for the beautiful Visitor Center.

Solar-powered pivots were built in the Mona fabrication shop, saving thousands of dollars of electricity.

Office and mechanic shop.

The demand for oil required expansion everywhere, and the farm became a very busy place. A new, gigantic 800 hp boiler was installed, and the old 600 hp boiler became the backup.

Gary continued to experiment with the distillation process and to modify his equipment. He even climbed into one of the extraction chambers with a flashlight and had them turn the steam on so that he could see how it moved against the dome lid of the distiller. He got the information he wanted; but needless to say, he didn't stay long in the chamber.

From what he saw, he immediately designed a new lid and again climbed back inside the cooker to see how it worked, but this time he poured water over his head to reduce the chance for burns. As he watched what happened with the steam, he was able to modify the lid for even better oil recovery. The GC analysis showed that nearly 15 percent more of the finer molecules were now being recovered than were being lost with the old design. "It's all about discovery so you do what you have to do to get the information you want," Gary said, laughing.

Dr. Hervé Casabianca came to see the new distillery and was impressed with the oil quality.

One of 12 greenhouses where at one time over 3 million starts were propagated that supplied both the Mona and St. Maries farms.

The second imported lavender harvester from France worked better, but it was still too slow as the plants had to be forked or dumped onto the trailer.

The third harvester imported the following year was more efficient. However, the plants were so big and bushy that the motor on the conveyor wasn't strong enough to carry them all the way to the top, so the men had to be there constantly to keep them moving. Elden always flew up from Texas to help, and he and Gary would take turns on the harvester.

The lavender is magnificent and when in bloom can be smelled miles away in the surrounding area.

Gary redesigned the French harvester and built a better one in the fabrication shop in Mona.

Peppermint is ready to be distilled.

Mary loved operating the swather. Hmm, stuck again? At least the aroma of the spearmint was refreshing.

Gary designed another conveyor system for harvesting peppermint and spearmint.

The first crop of clary sage was magnificent and produced beautiful oil, 1997.

Lush peppermint, August 1999.

Gary, Marta, Marcel, and Jean-Noël, holding Jacob, standing between the rows of baby lavender in St. Maries in 2002.

Marcel Espieu Comes to Utah

Finally, the time came to invite Marcel Espieu and his wife Marta to the convention to see what Gary had built. Marcel had often laughed at this American with big ideas, until the moment he arrived and saw what Gary had accomplished. Gary flew with them and Jean-Noël first to Idaho to see St. Maries. Marcel was amazed at the beautiful lavender and melissa and was thrilled to see the new baby lavender growing so well.

They flew back to Utah and drove down to the farm in Mona. Before the car had completely stopped, Marcel opened the door, jumped out with great excitement, and briskly walked through the lavender field touching and smelling the plants as he hurried toward the distillery.

At the 2002 Young Living Convention Farm Day, with tears in his eyes as he spoke to the group surrounding the distillery, he said, "The student has now become the teacher." It was the greatest compliment that Gary could have imagined.

Convention day at the farm each year hosts thousands of people as they explore the many facets of the farm, walk through the beautiful flowering fields, and marvel at the Seed to Seal process in such a peaceful environment.

The fields are stunning when in bloom with an aroma that wafts through the air for miles away. Hundreds of people come to participate in the harvest and/or the 5-K or 1-K runs through the lavender fields. Children and adults alike enjoy the happy and calming effects of lavender. Lavender essential oil is said to be the "mother of all oils."

A tour of the distillery is a must and is what the farm is all about. It is an awesome sight and a breathtaking moment when the oil drops bubble up from the condenser into the separator, and the "essence" of Seed to Seal is right before your eyes.

After the distillation has finished, the oil is taken to the decanting or filtering room. There is usually a little plant debris that escapes into the separator that has to be filtered out. The oil is also weighed and entered into the log book with other details specific to the distillation batch. When this is finished, the oil goes into large stainless steel containers that are taken to the warehouse. Samples are sent to the laboratory for analysis to make sure the compound structure is within the proper range, and when all the specifications are met, the oil is released for bottling and production.

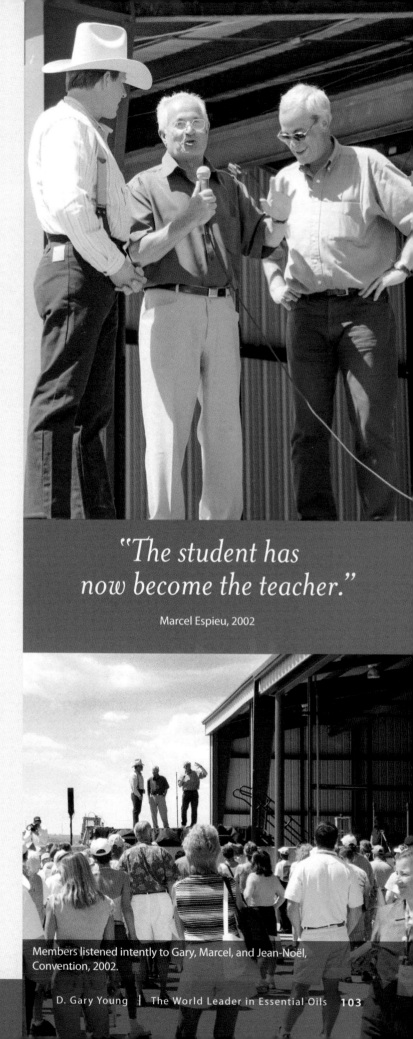

"The student has now become the teacher."

Marcel Espieu, 2002

Members listened intently to Gary, Marcel, and Jean-Noël, Convention, 2002.

Juniper is wildcrafted in the mountains south of Mona, Utah. The landowner is giving Young Living the trees in exchange for clearing the land to make more room for his cattle to graze.

An old tractor tire filled with concrete works very well to pack the cooker.

The beautiful red color appears when sunlight hits the junipe

German chamomile before the harvest.

German chamomile being loaded into the cooker.

The Mona Farm

Just as Gary envisioned in 1995, the Mona farm has become a world-wide destination for the entire family. The restored, old farmhouse is beautiful and quaint, offering an array of essential oil products and souvenirs.

One can't help but leave the farm with a tremendous appreciation for the drop of oil you are glad you have when you need it. The immense amount of time, effort, money, and dedication that is all part of Seed to Seal is indelibly imprinted in the mind forever.

The Percheron show barn is visited by thousands of people each year.

The medieval village hosts many events at the farm.

The medieval swing is fun and educational for the kids.

The 5-K run through the lavender has become a very popular event each year with people from all over the world.

The boys following in Dad's footsteps.

Grandma LaRue still yodeling at age 91, 2015.

Members and staff have fun together playing in the Western show. Everyone laughed when the town clown wanted to share his NingXia Red.

When Gary brought jousting to the farm, it became a very special and exciting event.

The chess tournament was a big success.

Gary says jousting is a challenge and loves all of it. The protective armor and chain mail weighs 90 to 100 pounds, and the lance weighs about 38 pounds. On a hot day, the temperature inside the armor is 5-10 degrees hotter than the outside temperature.

Bandido de Amores, the dancing Friesian, loves to entertain, making his trainer, Felix, very proud.

The eight-up Young Living world-class Percheron show horses are a thrill to watch. The show horses have taken many ribbons, and several hold individual first-place ribbons, including that of supreme champion.

Jacob and Josef kept the crowd laughing as they threw the javelin while riding their miniature horses, 2014.

Gary leads the procession for the opening ceremony with Felix and Jacob, while Mary sings the national anthem.

Freeman drives the Young Living Unicorn hitch in the competition.

The line up for the judging of the Eight-Horse Hitch.

The First Annual Draft Horse Show
Young Living Utah Farm, September 2015

Finally, the newest event took place at the Young Living Farm in September 2015 with the first rodeo, Draft Horse Show, and Einkorn Festival. The new arena is beautiful and spacious for the six-up and eight-up horse teams as they show their precision and beauty to the spectators. Draft horse teams came from all over the United States to compete and enjoy the environment of the peace and beauty at the farm.

The Lane Frost Challenge Bull Riding competition entertained everyone as the bulls snorted, the riders rode hard, and the clown distracted anything that was moving. An array of wagons, stagecoaches, "school bus" wagon, and penitentiary wagon took everyone back in time. The miniature horses pulling a miniature stagecoach were fun to watch as they bucked and kicked up their heels trying to feel their independence.

The event was magnificent and as many new people learned about aromatic crops and distillation, the excitement spread everywhere.

It was a marvelous event that promises to bring fun and excitement every year.

The dancing Friesians were exquisite in their performance and will certainly give the powerful jousting Percherons a challenge for taking "center stage." Different equine events will be a great addition to the farm, and there will certainly be plenty of manure to add to the composting for the natural fertilizer that feeds the crops.

Katie shows her beautiful Percheron in the Ladies Cart competition.

Rex Peterson, who has been training horses for the movie industry for 38 years, demonstrates with his horse, Tuff, that starred in such movies as *The Knick, Winter's Tail, Appaloosa,* etc.

Photosynthesis
HOW ARE ESSENTIAL OILS FORMED?

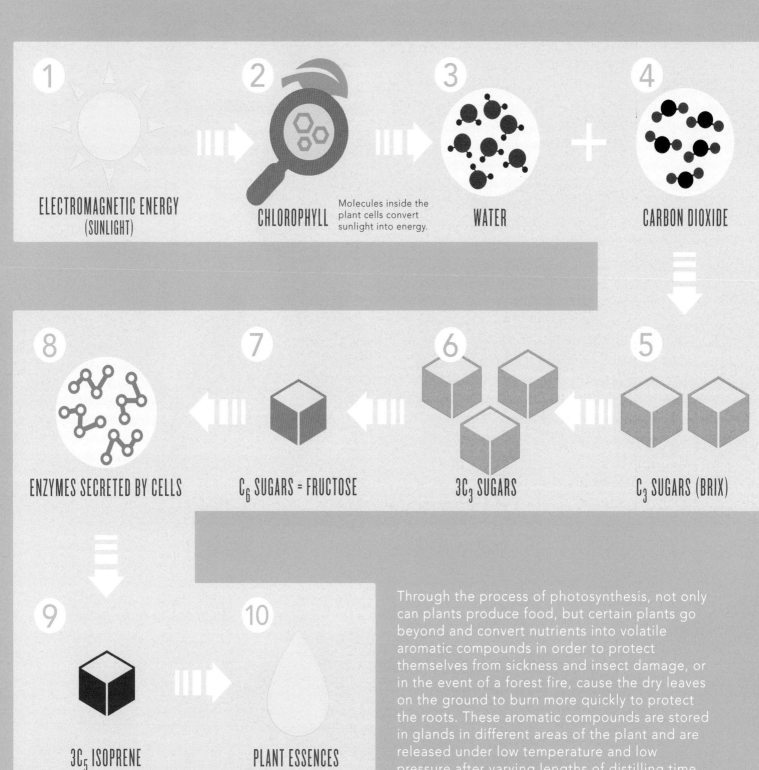

1 ELECTROMAGNETIC ENERGY (SUNLIGHT)

2 CHLOROPHYLL
Molecules inside the plant cells convert sunlight into energy.

3 WATER

4 CARBON DIOXIDE

8 ENZYMES SECRETED BY CELLS

7 C_6 SUGARS = FRUCTOSE

6 $3C_3$ SUGARS

5 C_3 SUGARS (BRIX)

9 $3C_5$ ISOPRENE

10 PLANT ESSENCES

Through the process of photosynthesis, not only can plants produce food, but certain plants go beyond and convert nutrients into volatile aromatic compounds in order to protect themselves from sickness and insect damage, or in the event of a forest fire, cause the dry leaves on the ground to burn more quickly to protect the roots. These aromatic compounds are stored in glands in different areas of the plant and are released under low temperature and low pressure after varying lengths of distilling time.

QUALITY AND PURITY

Gary's commitment to quality and purity begin with the Seed to Seal process. So what determines a genuine, pure, therapeutic-grade essential oil?

Every plant created by God contains DNA and messenger RNA (mRNA) (or what Gary likes to call memory RNA), which contains decoding instructions or the blueprint for the enzymes and sugars that produce essential oils.

So how are essential oils formed? In the process of photosynthesis, plants not only produce food; but in addition, some also convert nutrients into volatile aromatic compounds in order to protect themselves from sickness and insect damage or in the event of a forest fire, cause the dry leaves on the ground to burn more quickly to protect the roots.

The aromatic compounds or oils are stored in glands on plant stems and/or leaves. The essential oil is released under low temperature and low pressure after varying lengths of distilling time.

The Proper Environment

With the proper environment, soil nutrients, water, cultivation, and man's careful management, the highest quality essential oils can be produced that support the body's natural chemistry for health, spiritual awareness, and emotional stability.

The proper seed species is critical in order to produce the highest quality. But how do you know if you have the proper plant species unless you are the grower or working with the grower? What is the origin of your seed? How do you know that the seed is not genetically modified (GM)?

Most seed is commercially grown in a nursery not subject to the challenges of nature: sun, water, wind, and temperature changes. The harshness of nature develops strength in the plant's immune system that is expressed in the essential oil. Seeds grown in a controlled environment generally produce genetically weak plants, which usually produce a low-quality oil.

Growing healthy plants begins with understanding the physical and chemical properties of the soil along with the nutritional needs of the plants. Insufficient soil nutrients and poor soil conditions will result in poor essential oil quality.

Geographical location, soil type, climate, elevation, humidity, temperature, sunlight, frost-free days, rainfall, and many other variables determine the growth and health of the plants and thus the quality of the essential oil.

Aromatic plants grow best in well-drained soil rich in nitrogen, potash, and phosphorous with a pH of 7-8. Clay soil needs an abundance of compost, humus, enzymes, and microbes to break down its naturally rigid structure and change it to the ideal composition.

Composting

Composting increases the nitrogen, a very beneficial element in organic farming, which is practiced on Young Living farms worldwide. Discarded foods scraps such as banana peels, coconut husks, and skin peeled from mangos, papaya, apples, etc., are added to clover and alfalfa to begin fermentation. When all of this is mixed with manure and the liquid worm castings, we have an excellent fertilizer that provides a high source of nitrogen to feed the plants.

Weed Control

Weed and pest control are always major challenges in organic farming. Gary Young has developed a natural herbicide made with essential oils, Neem oil, and Castile oil. His pest control spray is a combination of Cinnamon, Palo Santo, Basil, Idaho Tansy, Pine, and Citronellal, with Neem oil as a carrier oil. Their efficacy is quite remarkable. It has been very interesting to watch how the different plants respond in a positive way to the essential oil spray.

Weed and pest control sprays made with essential oils can dramatically improve the organic environment of the plants. Soil sprayed with certain essential oils has been shown to digest unwanted chemicals in the soil as well. Naturally, it is much more expensive, but the results are well worth it.

Weeds can be a farmer's worst enemy and one of the biggest challenges when growing organically. For that reason it is important to control the weeds through cultivation early on as the plants develop.

1. Summer fallowing cuts the weed growth and prepares the ground for planting.

2. Planting cover crops such as beans, corn, triticale, clover, etc., help choke out the weeds and add more nutrients when it is plowed into the soil.

3. Rotating the crops is always beneficial and adds different nutrients to the soil.

4. Some aromatic crops like chamomile and clary sage grow so close together that they choke out weeds.

Organic Farming

Young Living practices organic farming methods and requests the same from all partner and cooperative farms. The Young Living farm in Ecuador is very rural and far from any industrial and city pollution. The air is fresh and uncontaminated, which lends itself very well for organic practices and certification.

In 2013 the Young Living Ecuador Farm received organic certification for the plants, raw farmland not yet planted, and for our cacao from CERES, which is a German organization accredited by the U.S. Department of Agriculture, comparable to USDA Organic certification.

A worker applies an essential oil-based spray to the plants.

Distilling Clary Sage

Gary had been distilling clary sage at the Mona farm for over an hour when the water-supply pipe to the boiler broke, shutting down the distilling. The oil yield was good and the distillation was almost finished, so the workers wanted to unload the cooker and prepare for the next one. But Gary had a thought, "Let the material sit in the cooker until we are ready to start again, and then let's distill for another hour and see what happens."

He was amazed when he discovered that during the second distillation period, there was greater extraction of the very desired compound sclareol. Allowing the clary sage to sit in the cooker until the pipe was repaired softened the fiber channels, which released even more sclareol. The oil from the two cooks was combined, achieving a greater balance of all of the constituents. The Young Living farm in France is now using this same distillation technique.

The environment greatly affects plant growth and the quality of the oil that's produced. Latitude and longitude are complex factors that determine the compounds produced in the plants.

Many questions can be asked about distilling each plant; but for each plant, there will be many different answers. Only someone like Gary Young, who grows, harvests, and distills plants in different countries with different climates and soil conditions, could give complete and in-depth answers.

Sometimes, it takes years of working in the same area on the same farm with different plants to answer some of these questions. Many people can go and visit a farm, watch the workers in the field, observe the entire distillation process, and think they know all about distilling; yet they couldn't duplicate what they saw or answer detailed questions about the operation.

Preparing the soil, planting, cultivating, distilling, and scientifically testing are all part of the education to understanding the whole process, besides the problems that come with climate changes and the solution for the problems that come with those changes.

Extensive research will always be an on-going part of Seed to Seal, including soil composition, climate and environmental statistics, planting, cultivating, harvesting, distilling, and analytical testing. Planting a new crop is very expensive and can be very costly if a wrong decision is made.

Helichrysum grew beautifully in Utah, but due to the high elevation and the cold winter temperatures, the neryl acetate compound was too low for Gary's approval. Even though it had a nice aroma, it was dug up and replaced by a different crop.

A truckload of clary sage arrives at the distillery.

WOLFBERRY

In 1993 Gary met Dr. Cyrus McKell, one of Mary's neighbors, who was the dean of the Botany Department at Weber State University. Dr. McKell had been doing some collaborative research with Dr. Songqiao Chao, a Chinese scientist and dean of the Science Department at the Beijing Technical University. Dr. McKell had invited Dr. Chao to lecture at Weber State University, and while in Utah, invited him to go to the Young Living office to meet Gary Young and introduce him to essential oils.

Dr. Chao was fascinated with the oils and wanted to know more. At the same time, he asked Gary if he knew about Ningxia wolfberries. Dr. Chao had directed a research project with his students on the *Lycium barbarum* wolfberry species growing on the Elbow Plateau in Inner Mongolia. It had been noted that the people living in the area who were eating large amounts of wolfberries seemed to have greater longevity than those not eating wolfberries. There are many species of wolfberries, but the *Lycium barbarum* produced the best results compared to other species in China that showed very minimal or almost no extraordinary benefits.

Dr. Chao's daughter, Sue Chao, a biochemist teaching at the University of Utah, subsequently came to work for Young Living to manage the laboratory. She translated her father's research and documentation into English and gave that material to Gary. He became very excited and wanted the wolfberries for Young Living. Gary found a grower of this particular species on the Elbow Plateau in the Ningxia Province of China and was the first to import tons of the berries for commercial use.

Those eating this superfood, as some call it, say they feel a greater flow of vital energy that supports their physical and emotional well-being. Wolfberries contain an abundant mixture of highly concentrated essential nutrients in perfect balance; and with essential oils added to the juice, they become even more desirable.

Testing high on the ORAC scale, in addition to all of the wonderful benefits, makes it easy to understand why wolfberries have been used and treasured in China for centuries.

More than 3,000 pickers work during the harvest time from June through October.

Dried wolfberries are very popular for making tea and in all kinds of cooking recipes for muffins and cookies, trail mix, and are a great combination with einkorn flakes. In China they are used as a garnish for many dishes.

Our wolfberries are processed in a very sophisticated and highly technical facility, where they are washed, dried, and packaged for international shipping.

Gary loves to create new recipes and had a lot of fun

Wolfberry puree is the foundation for our delicious NingXia Red.

Gary first found einkorn growing in Hunzaland in 1995.

EINKORN

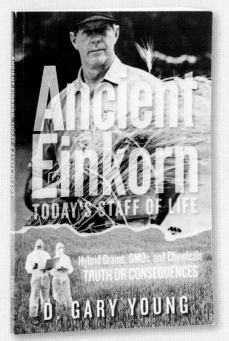

Today, on the Young Living Farm in France, Gary Young is planting nonhybridized einkorn seed and then harvesting it the old fashioned way by letting the cut stalks stand in the field for 7 to 10 days to allow the germination process to start before threshing.

When Gary began searching for this ancient grain back in 1990, he found it to be very illusive. He discovered small patches growing in Hunzaland, in remote areas of Turkey, and eventually on the east bank of the Jordan River Valley.

A few years later, Jean-Noël became interested in einkorn and talked to Gary about it. Neither of them knew a lot, but they wanted to know more, so Jean-Noël planted some for testing. Then one of our co-op farmers was engaged to plant some seed, which increased our productivity and produced enough seed so that Gary was able to bring home a little seed to test plant to see if it would grow in Utah. It did very well and produced more seed, and each subsequent year the amount increased to where Gary cleared enough acreage at the farm in 2012 to plant 150 acres of einkorn. Several farmers in the United States and Canada have been sent seed from the farm and are growing and experimenting to see how well the seed produces in their areas.

The history of wheat is very interesting, and much can be learned from the evolution of what some historians call the "ancient, original grain of man" to the hybridized "dwarf" wheat of today. As the population of the world increased, so did the need for more food. Wheat seemed to be a good choice that could feed millions of people. Scientists began looking for ways to increase the yield, which led to the hybridization not only of wheat but also of other grains as well.

The "new" wheat made the harvest easier and faster and reduced the cost to the farmers. However, the consequences of altering Mother Nature's wheat were not taken into consideration. Now there wasn't time to germinate or activate the enzymes. The fact that wheat became difficult to digest, rendering many nutrients unusable, and causing more and different digestive problems to appear were not results that seemed to bother anyone; or people simply didn't put two and two together.

As Gary continued to learn more about strengthening his own body, he noticed these things. His analytical mind kept asking questions, which took him down this path of discovery. Going back to nature—to the original grain—seemed to be the thing to do, which started him on his quest for einkorn. The more he studied, the more certain he became that this was an ancient food that needed to come back to the modern world. How amazing it is to think that this realization was so similar to the ancient knowledge of the essential oils he wanted to research and restore to mankind. For more information on Gary's book, visit DiscoverLSP.com.

Gary is passionate about growing the ancient grain of einkorn and planted the first test crop in Mona in 2011.

The early 1900 steam engine powers this vintage thresher. It is a rare moment to go back in history and see the old equipment in operation.

Gary became driven in his quest; and because of that "knowing" that drives him, einkorn flour ground from the harvested kernels first made its way into the Young Living market as pancake mix and spaghetti. Now the flour has become available, and people everywhere are experimenting with all kinds of recipes using the flour as well as the pancake mix. An einkorn mixture dipped in chocolate and sold as Einkorn Nuggets has been a fun creation, and a breakfast cereal or snack with kernels pressed into flakes and mixed with nuts and wolfberries will soon enter the market.

All of this comes to us because Gary had the vision that he knew would benefit so many and give an alternative to a vast number of products made with hybridized grains. It would seem that if this rediscovered einkorn grain produces the high nutritional value as it did anciently, then it would make sense that we could eat it and benefit from it in the same way the early people of our world did; and we certainly don't have to travel to the Middle East to get some.

Scientists who have been studying einkorn for years say that eating nonhybridized einkorn grain means that we are eating nonhybridized gluten. Therefore, einkorn is easier to digest, contains vitamins, minerals, and proteins that are important to our health, and is probably the most preferred grain to eat. It also has a delicious taste.

Gluten-free fad diets, which have become popular today, will eventually cause challenges down the road. Our bodies need and want what comes naturally in our food that has not been altered genetically. The body knows what to do with what God has given us to sustain life and should respond in a normal and positive way.

Due to the increased volume of kernels on the stalk as a result of hybridization, scientists thought that more food could be produced and more people could be fed. This seemed like a great solution for feeding the people of world. However, when the genes of a plant are altered, the God-given life force is changed within the plant.

Einkorn grown at the Mona farm is left in the field for a week, so the kernels can begin germinating before being threshed.

Einkorn on the left and modern hybridized wheat on the right.

Harvesting einkorn in Utah, 2013.

Dwarf wheat and other modern grains contain 42 chromosomes, with more and different varieties of genes for gluten proteins, causing possible body malfunctions. Other grains such as spelt (often called kamut) and emmer have 28 chromosomes, which is less than the more modern grains, but is still more than einkorn.

Hybridized wheat with 42 chromosomes creates different genetic codes for new proteins that man was never meant to consume. With 3 genomes (A, B, and D), it is the hybridized wheat D genome that is the source of gluten-triggered responses. The benefits of changing the genetic structure of a plant for whatever reason does not compensate for the negative effects.

The differences in the kernels of grain are easy to see. The einkorn stalks are long and slender, and the kernels are oblong and flat-looking, protected by a strong husk or chaff that clings to the kernels and is more difficult to husk. Because the einkorn hull is difficult to remove, it protects the kernels against negative environmental factors and destructive pests.

Because einkorn has the simplest genetic code of all varieties of wheat, with just 1 genome and 14 chromosomes, it is easy for the body to utilize.

Nonirrigated or wild einkorn grows about 2 feet tall. In wetter climates and/or with irrigation, the cultivated einkorn grows 4.5 to 5 feet tall with hair-like tassels that wave when the wind blows.

Hybrid wheat has been engineered to grow between 14 and 16 inches high with shorter, thicker stalks to support the heavier grain heads. The common hybridized "dwarf wheat" of today has double the number of kernels that are thicker and heavier with less hair and a soft husk that is easy to remove, making it more susceptible to pests and various fungal diseases.

The kernels of the fully matured hybridized wheat fall to the ground as soon as the harvesting equipment begins to shake the stalks. For years, farmers have been harvesting the wheat green in the soft dough stage to capture the kernels before this happened.

Hybridized wheat enables modern harvesting to bypass the different stages of development. Today, the grain is cut and threshed immediately, so it has no time to go through the maturing stage to germinate and develop the enzymatic activity needed later for digestion.

The modern way wheat is harvested is for man's convenience, not for maximizing health benefits. In ancient times farmers would not have understood the science of harvesting wheat, but they must have known that the wheat was more health-sustaining when it went through all the stages of maturing.

Gary harvesting einkorn on the Young Living farm in France, 2014.

Distillery site

This is what Mary and everyone else saw as they looked at the dense jungle, November 2006.

This is what Gary saw as he looked out over the immense jungle and saw what it would become, July 2012.

ECUADOR—A NEW OPPORTUNITY

In 2005 Gary's path took another unexpected turn when he was invited to Ecuador by a university professor, who was the dean of the medical school in Cuenca. He had an interest in essential oils and wanted Gary to develop an essential oil program with various research projects for his students. However, it didn't take long to discover that the university administration didn't have a lot of knowledge about the prerequisites and curriculum that would be necessary for this type of study in essential oils, and they were not prepared to offer such a course.

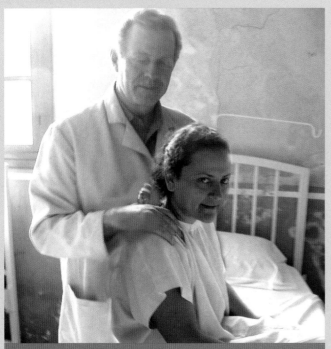

Gary really enjoyed the Cinterandes mobile medical team when he was invited to travel with them into the jungle to conduct a research project using essential oils.

Gary did have the opportunity, though, to travel into the jungle with the volunteer Cinterandes medical team to conduct research with essential oils. He loved the opportunity to teach more about essential oils and to become acquainted with Ecuador and its vast possibilities with its many different plants and eco systems.

Was Gary disappointed with the university? Perhaps, but then maybe he was relieved. He was in Ecuador, a country with an unmatched ecosystem with diverse climates and terrain and thousands of unidentified plants and trees and so much not yet explored. He quickly realized that the climate in Guayaquil would enable him to farm year round, which was very exciting; and there were discoveries to be made in the untapped world of essential oils.

Finca Botanica—Young Living Ecuador

The challenge of finding land was arduous and costly; and as inexperienced Americans, Gary and his family were often the targets of unscrupulous individuals. But Gary was determined and no matter what happened, he was going to continue on the path he had chosen.

In September of 2006, he found the perfect place—raw, dense, unwanted jungle on the edge of the remote town of Chongon about 40 minutes from the Guayaquil airport.

The task seemed overwhelming; but with indescribable perseverance, against the odds of unrelenting mosquitoes, stinging insects, no-see-ums, scorpions, boa constrictors, poisonous snakes, and monsoon rains that turned the ground into glue-like mud, almost impossible to traverse, 2,300 acres became the home of the next Young Living farm.

In September 2006, Gary found the perfect place and in November, on Thanksgiving Day, the papers were signed for the land, and the transformation of the jungle began. The soil was dark and rich with nutrients but thick with clay, which took a lot of time with heavy equipment to move.

Gary, operating the D8 bulldozer, cut the first road, built the dam, and carved the beginning out of harsh, thorny bushes, shrubs, and trees. When the monsoon rains came, the valley quickly filled with water ready for thirsty crops in the tropical heat. The farm evolved into a working "metropolis," hidden from the world in the dense jungle.

Gary built 2½ miles of road into the farm.

It was a remarkable time and probably no other man in the world would have undertaken such a project; but the dream was vivid, his determination unstoppable, and the constant discovery of new plants filled Gary's head with unending ideas.

Mary's "first" house.

The first greenhouses were built next to the dam.

Walls were framed with bamboo, and the concrete was mixed by hand.

Gary had the 350 hp boiler built in Guayaquil in 2007 and shipped parts and equipment from Utah that could not be found in Ecuador.

The 14,500-liter cooker was ready to go to the farm.

Containers arrived from Utah with needed building materials and distillery parts.

The battle of the monsoons was often frightening; but with Gary, it was all about how to solve the problem, not to let the problem stop him.

The torrential monsoon rains came down like sheets of water that were difficult to penetrate or see through. The culverts weren't big enough and the road was washed out. The clay soil turned to sticky mud and was as slick as an icy winter road.

During the monsoons, the water could rise 1 to 2 feet per hour. This small break in the dam could have been a disaster, but Gary was there and everyone moved quickly to plug the opening and stop the rushing water.

The pictures are never ending, but Gary and John love it.

The road leaving the farm between the second dam and the field of dorado azul is the main thoroughfare through the farm. Over 6 miles of electrical wiring were strung from the main road to the farm.

The Palo Santo Forest is a special place to visit.

Palo Santo is a very unique oil that is extracted only from the distillation of dead wood that has been lying on the ground for years. Gary learned about this wood in 2006 while traveling throughout Ecuador and had heard about the palo santo oil from various people. He finally found an elderly gentleman who had been distilling small amounts of wood and selling the oil locally. Gary was intrigued with the oil and took enough wood with him back to Cuenca to distill in his small testing distiller. He sent the oil to Utah to be analyzed; and when the report came back, he decided this would be a wonderful addition for Young Living.

So, when he moved to Guayaquil and set up the small distilling operation before the farm was established, he began looking for villagers who could gather the dead wood for him. He learned that the longer the tree had been dead, the more oil was found within the channels of the wood fiber, a phenomenon known only to Mother Nature. The people in the remote jungle villages began to gather the wood and truck it to the farm.

The Palo Santo Forest at the farm covers over 100 acres of undisturbed peace and quiet.

Prior to this, the villagers had no opportunity to work, so this wildcrafting for Young Living has greatly improved their lives. With the money earned, they have been able to build two schools. The supply of wildcrafted plants will always be limited and without guaranteed availability, but those who have this oil know of the deep feeling it gives of being grounded and close to nature.

When Gary was clearing the land to prepare the new fields for planting, he discovered thousands of palo santo trees growing in the brush on the property. He sent the farm crew to walk ahead of the bulldozer to safeguard the trees until they could be moved.

Earlier, Gary found an area deep within the farm that had an abundance of palo santo trees. It seemed like a natural sanctuary and was designated as a special place. Over 3,000 trees were transplanted to this new location now known as the Palo Santo Forest, which covers more than 100 acres of high rolling hills. The Palo Santo Forest is a quiet place for contemplation, meditation, and spiritual renewal and is enjoyed by everyone who comes to the farm and takes time to go there.

Dead palo santo wood is gathered and taken to the farm to be distilled.

A small patio next to the house that Gary leased for the family was enclosed for experimental distilling of ruta and palo santo. Ruta was first distilled in 2007.

In the highlands near Pelieo and Ambato, about 500 people grow ruta for Young Living.

Gary built employee housing on the farm with electricity and hot and cold running water, a first for most of the workers, as well as a washer and dryer next to their housing. It was amazing to watch as none of them had ever used a washing machine.

Hundreds of people have jobs for the first time, and altogether, Young Living is providing an income for over two thousand families throughout Ecuador. People around the world are blessed because of the oils produced at this farm.

He made eight expeditions into the Ecuador Amazon jungle on foot and by canoe, speedboat, and whatever means was available. Even at the risk of his life, he continued to explore and find new plants, which he brought out of the jungle to distill and analyze. When they met his specifications, he engaged the local people to harvest the plants or flowers and bring them to the farm, providing work and income for natives who had never had an income before this time. Many were anxious for the opportunity; and after several years, they still bring the plants to the farm to be distilled, enabling Young Living to offer unusual, exotic, and beneficial oils that are part of the Young Living essential oil line.

Employee housing built on the farm has four individual apartments, each with three bedrooms and two bathrooms.

Lush vegetation grows in nutrient-rich soil produced by the many volcanoes of Ecuador.

Gary is always looking for new plants.

The guide had never seen turkey hot dogs before but ate them with gusto.

Visiting with children in the Amazon rain forest. They loved Wolfberry Crisp bars.

Gary explored the Amazon by longboat.

Gary discovered the unidentified plant that he named dorado azul. It was considered a jungle weed that had no value, but it had a beautiful smell; so Gary test distilled it and decided to cultivate it on the farm. It grew well during the monsoons and was the first major crop distilled at the new farm in 2007. It has since become a favorite oil in Young Living.

Dorado Azul (*Dorado azul guayfolius officinalis*) was discovered growing wild in the foothills northeast of Guayaquil. Because it had no identification, Gary named it, harvested it, and distilled it for the first time in 2007. Dorado Azul thrives in the monsoon rains, which would drowned out so many other crops. After evaluating the oil, Gary went back to the brushy jungle, gathered the seed, and started the first field domestication of that particular species.

Dorado azul has four other related species that look similar but with slight leaf and flower variations and also different oil chemistry. One of the different species was discovered in 1806 and named *Hyptis suaveolens*, according to Wikipedia—http://en.wikipedia.org/wiki/Hyptis suaveolens—and can be found today in the foothills from Melvis in southern Ecuador to the border town of Tumbus, Peru. The variations of this plant are similar to lavender (*Lavandula angustifolia*) and spike lavender (*Lavandula latifolia*).

Gary built Mary's "first house" on the far side of the dam, but she declined because of the snakes, spiders, scorpions, and other creepy crawlers. The second farmhouse was much more appealing.

Ruta was the first crop to be distilled, 2007.

In August of 2007, the first distillation took place—a remarkable feat; and those who visit stand in awe of the accomplishment. Gary carved this beautiful, producing farm with all of its attractions out of unusable, wasted jungle land.

Three 14,500-liter extraction chambers distill large quantities of plant material 6 days a week, 11½ months of the year; and during some harvests, the distillery operation runs 24/7 to extract the oils at exactly the right time. The glucose content (Brix measurement) of the plant is monitored closely to ensure that harvesting and distillation happen at the right time. Distilling even one day too soon or one day too late can reduce the amount and quality of the oil produced.

350 hp wetback boiler built in Guayaquil.

The shop is large enough to work on heavy equipment inside during the monsoon season. The office is on the second floor.

A 6-ton block of distilled plant material is lifted from the cooker with a 10-ton hoist.

Everybody helps; early training.

Critical to the extraction process is the cooling system that runs through the condenser. Because of Ecuador's hot climate, it was a challenge to figure out how to cool the water without shipping cooling units from the United States for a cost of about USD 250,000 each. That wasn't a feasible option, so Gary became very creative and came up with the idea to put copper tubing in an ordinary deep freezer and fill it with anti-freeze to keep the pipes cold. It seemed ridiculous and many people laughed, but when it was done and all hooked up, it worked beautifully and has worked extremely well since 2011.

On the lower level of the distillery are four different smaller sizes of distillers that are used for the extraction of oils such as Ylang Ylang, Palo Santo, new crops and plant material for testing, and crops that need to be distilled in smaller batches.

These smaller chambers are used for research when new plants are found and for crops brought in by the co-op farmers and natives who wildcraft in unfarmable regions. Some of the natives from the jungle are earning an income for the first time with the crops they bring to the farm. New plants are constantly being discovered; and their oils are distilled and analyzed, and some become new oils for Young Living.

Four small distillers are used for testing new plants and crops that need to be distilled in small quantities.

In tropical climates it becomes a challenge to cool the water enough to recover all the oil from the distillation. If the water is not 50ºF or colder, there will not be adequate separation of the oil and water. The idea came to Gary to use a deep freezer as a heat exchanger. He bought one for each condenser (against the wall) that was modified with copper tubing and filled with antifreeze to keep the water cool to complete the process.

Two more 14,500-liter cookers were added to meet the growing demand, and flex tubing was imported from the U.S. for Gary's modified design.

PRODUCTO TERMINADO
FINISHED PRODUCT

Gary teaches the members about the distillation of dorado azul.

In June 2011, land was cleared for 24,000 ylang ylang trees. Members planted thousands of seedlings; and by June 2013, the trees had grown into very big, bushy trees that produce flowers year round.

Nicolas, the farm manager on the right in the white shirt, checks to see how the ylang ylang harvest is progressing. Francisco, field manager (nicknamed Panchito, on the left in the green shirt), was the first local person from Chongon to be hired.

Happy ylang ylang pickers. Primo (in front) was transferred from St. Maries to Ecuador because of his farming and distillation expertise that he acquired after working for many years on the St. Maries farm.

The original 24,000 ylang ylang trees expanded to 48,000 in 2013.

The smell of ylang ylang is intoxicating.

Ylang ylang flowers are sorted and distilled in the small distillers on the lower level of the distillery.

The restaurant, built with bamboo, serves over 250 nutritious meals every day.

Farm workers, spa staff, and many guests enjoy nutritious meals prepared every day in the kitchen.

During the monsoon season, the reservoir level comes up so high that it creates the appearance that the restaurant is floating.

The restaurant that sits on stilts over the reservoir serves over 250 nutritious meals every day to 150 farm workers, spa staff, clients, and many guests who come to visit. Acres of rice are grown on the farm for the restaurant and the school. The garden grows numerous types of vegetables and fruit galore, especially with the banana and mango plantations that are within a two-minute walk from the kitchen. Special meals are prepared for those in need at the spa; and all the meals are organic, with no sugar, white flour, or other unhealthy ingredients. The tropical fruit juices made daily are organic and delicious. Goat cheese and yogurt are made from the organic milk that comes from the goat dairy on the farm. It is a tourist's delight for all those who come to visit.

Vallorie, Tamara, and Mary love the sweet bananas.

Nothing compares to the delicious taste of organic mangoes that have been ripened on the tree.

The amazing yacon is a potato-like vegetable that produces a sweet, mineral-rich liquid when pressed.

One entire hillside at the farm is covered with thriving Sasha cocoa trees from which Young Living organic chocolate is produced. Ecuadorean Sasha trees are the favored trees because the beans have a delicious flavor and are very productive. One hectare (2.47 acres with 1,100 trees) of Sasha trees can produce 2-3 tons of cacao beans per year.

Eucalyptus blue is gathered in the eastern Amazon region of Ecuador and brought to the farm for distillation.

EUCALYPTUS BLUE

Components	Accepted Lot 26808	Rejected Lot 24921
Alpha-Pinene	19.9	28.9
Beta-Pinene	0.7	0.4
Myrcene	0.7	0.7
Alpha-Phellandrene	0.4	0.9
Limonene	5.4	5.4
Eucalyptol	66.2	4.1
Gamma-Terpinene	0.6	1.2
Alpha-Terpineol	0.6	1.8
Alpha Terpinyl Acetate	0.8	3.8
Aromandendrene	1.6	16.7

These are the major components that make up the action of the oil. Minor components have been left out.

Workers harvest 100 acres of lemongrass, which produce four crops yearly.

Gary found the oregano species *Plectranthus amboinicus* growing on the farm in 2008 and distilled it in 2009. After the analysis, he determined that this was the very best oregano of the future and has expanded the fields to 100 acres.

Oregano in the greenhouse will soon be transplanted into the fields.

PLECTRANTHUS OREGANO (ECUADOR OREGANO)

Constituent	Area % 012706	French Medical Encyclopedia ORIGANUM VULGARE
Alpha Thujene	0.70	
Alpha Pinene	0.40	
Myrcene	1.90	
Alpha Terpinene	12.20	
Para Cymene	3.80	
Eucalyptol	0.40	
Gamma Terpinene	17.90	25
Terpinen-4-ol	1.20	
Alpha Terpineol	0.60	<10
Pulegone	0.50	
geraniol	0.30	
Geranial	0.60	
Carvacrol	32.90	60-70
Trans Beta Caryophyllene	10.20	
Trans Alpha Bergamotene	6.70	
Trans Beta Farnesene	0.30	
Alpha Humulene	2.80	
Beta Bisabolene	0.40	
Delta Cadinene	0.30	
Caryophyllene Oxide	1.7	

Worm Houses/Worm Castings

Millions of California red worms multiply in six worm houses on the Ecuador farm, creating liquid fertilizer and castings (manure), which are applied every week on the fields on a rotation basis and mixed into the potting soil for the greenhouses. The liquid produced from the composting is mixed with the liquid from the worm castings and goes through the sprinkler system for irrigating the crops. It is a unique way for controlling automated fertilization and has proven to work very well.

Holding tank for liquid worm castings that is mixed with compost and goat manure.

Manure is trucked in from nearby dairies and used for soil enhancement and composting.

Composting is a key element in organic farming on all Young Living farms and is a beneficial way to use the plant material or "straw" after distillation, which is mixed with discarded fruit and vegetable scraps and manure from the goat dairy. In Ecuador over 5 million tons are spread on the fields annually.

The goats produce organic milk and cream for the restaurant and the Young Living Academy. The manure is mixed with the compost for the fertilizer that goes back on the fields.

Research Nursery

In the nursery we record plant and root growth with different soil conditions, adaptability, pest resistance, and weed control with oils. We also study new plants found in the jungle to determine their nutritional value and water needs. When ready, seedlings are transplanted into the fields and studied for their growth and potential oil production. Chris and Gary check to see if the plants are ready to be transplanted.

Seeds are germinated year round and are continually transplanted to the fields.

This sophisticated weather station constantly monitors weather patterns, UV hours per day, temperatures, humidity, barometric pressures, rainfall, and daylight hours. This data is used with the Brix testing to determine the best harvesting and distilling times to maximize production.

Examples of Brix testing:			
Lavender	St. Maries	Shade dried for 62 hours	Brix 24
Lavender	Mona	Shade dried for 48 hours	Brix 28
Lavender	France	Cut before the flower goes to seed, dried 76 hours	Brix 21
Melissa	St. Maries	Cut mid bloom, shade dried for 12 hours	Brix 14
Oregano	Ecuador	Cut before going to seed, shade dried 120 hours	Brix 24
Dorado Azul	Ecuador	Distilled immediately	Brix 16
Peppermint	Mona	Cut in full bloom, sun dried for 3 days	Brix 28

Brix Testing and Harvesting

What is the perfect time for harvesting? Many variables determine when the plants are ready to be harvested, beginning with the weather.

1. A growing season that has had daily low temperatures or high temperatures, heavy rains, or little rain will greatly affect the maturing time of the plants.

2. As harvest time approaches, multiple daily Brix-level testing of the plants, which is the measurement of the plants' glucose (sugar) content, helps determine the time of day to harvest. The higher the Brix levels, the greater the oil production and the higher the percentage of chemical compounds.

3. When the plants are ready to be harvested, a small amount should be cut for a sample distillation.

4. The distilled oil is then scientifically tested with a GC/MS instrument to determine its chemical profile, which indicates constituents and their percentages for that particular oil. If the chemical profile does not meet Young Living's standard, then either more maturing time is needed, or the crop will not be distilled. This is one reason why new plants are test distilled to see if that particular plant has enough benefits to merit cultivation or further wildcrafting.

Only one drop of plant juice is needed to measure the Brix with the refractometer.

Curing is a very interesting and critical factor of distillation. Gary discovered in his early years of distilling that there was a difference in the volume produced and compounds found in the oil of each batch of plant material, depending on how soon it was distilled after being harvested. Some plants are very stressed when they are cut and will manufacture more oil within the plant fiber as a way of trying to protect themselves.

Gary did not take this into consideration until he started distilling in different states and then in different countries. It was very revealing to see how the distilling was different in Washington, Idaho, Utah, and then France, Ecuador, Oman, Egypt, Taiwan, Israel, northern Canada, Croatia, and other areas where we own or manage our partner farms.

It was interesting to discover that geographical location made a significant difference in the distillation. North and south of the equator and the 45th parallel, elevation affects the maturing time of the crops, which may require a modification of the distillation process in order to obtain the highest quality oil with the best constituent profile.

North of the 45th parallel, we have Washington and Idaho, but Utah is south of the 45th parallel. Then the elevation adds another factor to consider. Washington, north of the 45th parallel, is at an elevation of 1,500 to 2,500 feet; and the time for distilling peppermint is early August.

Below the 45th parallel and above 4,500 feet, in order to have higher levels of menthol and lower levels of menthone, the best time for distilling peppermint is late August to early September.

In the hot, humid climate of Ecuador, because we can farm all year round, there is more time to test and conduct research to determine the best distilling time.

Shade drying and maturing the plants before distilling varies greatly with different plants, even in the same geographical area, as you can see from the chart.

Because shade drying is so important for so many crops, Gary extended the concrete pad and roof to protect the crops from the direct sunlight to ensure the proper curing time.

A large number of factors must be considered during distillation in order to obtain the highest quality oil.

Plant Material	Distilling Time
Dorado Azul	Immediately after cutting
Ruta	5 days after cutting
Oregano	5 days after cutting
Lemongrass	3 days after cutting
Vetiver	7 to 10 days after digging
Eucalyptus Blue	3 to 5 days after picking
Ylang Ylang	Immediately after picking
Ocotea	10 days after picking
Palo Santo	5 years after the death of the tree

Only someone who grows, harvests, distills, conducts analytical tests, and keeps records would know these details.

Too little sunshine, too much rain, or a dramatic change in temperature can change the distillation time, causing it to vary from year to year. A lot of experience is needed to understand the distillation of so many different plants with so many complex possibilities. Different plants may grow in many different types of soil, but will these plants produce oil?

Many questions need to be asked:

1. What is the condition of the soil, and what is its nutritional profile?

2. Can lavender grow in a 5.5 pH sandy or clay loam soil?

3. Can frereana frankincense grow in Oman?

4. Will ylang ylang grow in Idaho?

5. Will lavender grow on the tropical coast of Ecuador?

Many of these questions are constantly asked; but will these aromatic crops, planted outside of their normal growing environment, produce oil? Then even if the plants produce oil, will the oil have the right chemical profile with the right percentages of the compounds?

The answer is, "No, they won't, or not usually." Essential oil compounds are produced from the plant's genetic profile and fed by the nutrients in the soil. If the right combinations of nutrients are not in the soil as well as the proper amounts of sunshine and rain, the oil compounds will not be the same.

Eugenio, who has been the distillery manager for nine years, checks the Brix levels of the plants every two hours as they cure before being distilled.

Ocotea leaves start to cure the moment they are picked and continue to cure and produce more oil during the 10-day journey from the jungle to the farm. When the trucks arrive, the leaves are spread out and turned twice a day until dry, which could take up to a week, depending on how wet they are when they arrive. It is critical that the material is dry when loaded into the extraction chamber because when the steam enters, the temperature of any moisture remaining on the plant material will cause the steam to prematurely condense, homogenizing the finer molecules of the oil, compromising their extraction and recovery.

Isaias spreads lemongrass for shade drying and curing before distillation.

In Gary's own words: "For 60 years, I have been working in the farming industry. I saw a lot growing up on a ranch where we raised oats, barley, wheat, and alfalfa at an elevation of 6,500 feet in central Idaho. My father tried to grow different species of alfalfa such as Ranger alfalfa, but it did not grow well in our environment or at our elevation. Besides that, it was quite common for the winter temperatures to drop to -20°F to -35°F for two to three weeks at a time. However, Ladak alfalfa grew very well and was very hardy.

"It seems logical that plants would grow well in similar climates; but when they don't, it is usually because the soil nutrient profile is different. When the aquifer is different, the subsoil water may be too much or too little for aromatic plants; and if that cannot be controlled, there will be problems.

"France has never in history irrigated lavender until 2015, but the climate has changed with less rain and higher temperatures. Many farmers have given up and now only 30 percent of the lavender growers in France are growing lavender today. The rest have switched to other crops.

"When I consider planting a crop, I first evaluate the growing conditions. If the soil profile is right, then longitude and latitude elevations have to be considered because this will change the planting and harvesting time, which in turn can change the quality of the plants. The number of warm days, cool nights, and UV hours of daylight are all variables that can change the molecular components of the oil, even the ketones of the aroma. Water is also a consideration. Too little water will kill melissa, and too much water will kill lavender or helichrysum, etc.

"We may think that the origin of the plant is the best place to grow it. However, I have seen lavender grow better in places other than France, but France is not the origin of lavender.

"The molecular structure of lavender in Utah is almost identical to France, but the harvested volume per acre is greater in Utah, which means greater oil volume per acre, thus increasing the revenue for the farmer. Because I am irrigating our lavender in France, other farmers are following and lavender is growing where water is available."

It is a remarkable feat, against all odds, to see this beautiful, producing farm in the surroundings of a dense and almost impenetrable jungle.

There was no water on the farm in the beginning. After Gary built the dam, the monsoon rains filled the reservoir, and the water spread out quickly.

This very sophisticated laboratory was built above the distillery in 2010, making it possible to have instant analysis during and after distilling.

The only GC/FID/MS combined instrument in the world.

GC/MS Analytical Instruments Become One—The Only One in the World

On the upper level of the distillery building is a full-scale laboratory where each distilled batch of oil is tested for chemical composition and overall quality. Verifying the chemical profile of an essential oil is critical to determining its value and best usage.

The gas chromatograph at the farm is combined with a mass spectrometer (GC/MS), giving the operator the best piece of equipment for separating and identifying chemical compounds. The GC/MS along with an optical polarimeter, which is designed to measure chirality, are highly sophisticated instruments that give the lab in Ecuador very detailed information about each oil.

This critical information validates Gary's decisions for the proper time for harvesting and drying, the best distilling temperatures, and the timing for each distillation. The purity and quality of the oils are extremely important in determining whether to go into full production of any given oil.

Gary initiated the gas chromatograph (GC) in Ecuador with dual flame ionization detectors (FID) and a two-capillary column system—one with a polar stationary phase (60m long) and the other with a non-polar stationary phase (50m long). The two capillary columns were connected to a splitter at the inlet of the GC so that the injection of one essential oil sample was evenly divided into the two columns. With this GC-dual column system, every component in nearly all essential oils could be separated and tentatively identified using Young Living's exclusive retention index library.

After the first year of operation, Gary added the mass spectrometer (MS) to this GC system, making it even more powerful. A 60m-long, non-polar capillary column was connected to the GC/MS portion of the instrument. No other GC/MS in the world is equipped with a total of three capillary columns running to two FID detectors and one MS.

The mass spectrometer allows Gary to perform research on new essential oils and new aromatic plant discoveries. The MS helps to confirm the identification provided by the retention index library. The Agilent mass spectrometer was made more powerful with the addition of a 500,000-component reference library.

The GC/MS operator simply selects the mass spectrum of an unidentified component, double-clicks, and the GC/MS computer searches for the best match in the 500,000-spectral reference library. In less than one second, it provides the top matches with the suggested name of the component. If the component name matches the retention index library value, then the identification of the component can be securely fixed.

In other words, there is one 50-meter and one 60-meter column for the GC and one 60-meter column for the MS. With the three columns, we are able to identify more oil compounds for a more detailed analysis. Gary wanted to combine the GC and MS instruments into one instrument. The manufacturer said it couldn't be done; but when he did it, they voided his warranty. Now, five years later (2015), the instrument is still working perfectly.

Dr. Casabianca (between Gary and Chris) loves the farm in Ecuador and is very impressed with the analytical laboratory.

Dorado Azul Chromatograms(s)

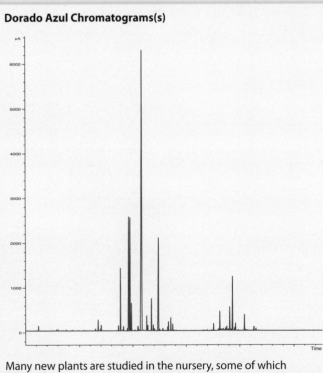

Dorado Azul

Components	Dorado Azul Dry	Dorado Azul Green	YL Specification	
	June 5, 2014 DA#001	June 6, 2013 DA#001	Min	Max
Sabinene	8.6	9.1	0.6	17.4
Beta-Pinene	5.9	6.1	3.2	10.8
Eucalyptol	33.8	36.8	28.4	54.0
Limonene			0.2	7.0
Alpha-Fenchol	14.1	13.4	2.6	19.1
Bicyclogermacrene	7.7	7.1	Trace	9.4

Other Major Peaks				
Cis-3-Hexene-1-ol	0.5	0.6		
Alpha Pinene	2.9	3.1		
3-Octanol	0.1	0.5		
Myrcene	1.2	0.8		
Fenchone	2.9	2.9		
Terpinen-4-ol	1.2	0.8		
Alpha-Terpineol	0.6	0.5		
Beta-Caryophyllene	2.5	2.1		
Germacrene-D	2.4	2.2		
Spathulenol	2.1	1.9		
Tau-Cadinol	0.8	0.7		
Tau-Muurolol	0.4	0.4		

Many new plants are studied in the nursery, some of which become new crops for producing exotic oils for Young Living.

The main difference between the green and dry samples is that there is a slightly increased amount of monoterpenes in the green sample over the dry sample.

Osias prepares special meals served in the spa.

The small restaurant initially built for the spa serves freshly made tropical fruit drinks all day long to everyone.

Gary built this tree house for the boys without a single nail piercing the majestic sabal tree. Many who come to stay at the spa have a lot of fun enjoying the comforts of home with two bedrooms, a bathroom, kitchen, living room, electricity, hot and cold water, and even the Internet—an adventure high in the jungle treetops.

NovaVita Spa and Rejuvenation Center

In 2006 Gary opened a research center in Guayaquil to test the efficacy of the oils. Clients came from all over the world desiring to be part of that research. New discoveries were almost a daily experience, and a great many people returned home with renewed health and vitality. However, the noise and pollution of the big city were not conducive to the well-being of everyone there, and so talk began about moving the center to the farm. Many people wondered how that was going to work, as it seemed almost impossible so far out of the city. But in 2012 the building went up; and when the construction was finished, the move began and the farm had a new attraction for visitors.

The NovaVita Spa and Rejuvenation Center sits in the middle of the farm, a stone's throw from the distillery. People from all over the world come to take advantage of the fabulous uses of essential oils. You can enjoy the infrared sauna, massage, Raindrop, even a chocolate masque with essential oils, besides the many other applications for rejuvenation. Many people come for just a simple "tune-up" and to enjoy the peaceful atmosphere of the farm. The essential floral water from the distillation, which would normally be recirculated or used for irrigation, is piped underground to the spa, where it is stored in huge vats to be used in the hot tubs.

Both clients and visitors can enjoy a relaxing and rejuvenating time as their skin soaks up the micro oil molecules from the hydrosol that are not separated during distillation. Imagine soaking in Palo Santo, Dorado Azul, Eucalyptus Blue, or even in the exotic Ylang Ylang floral water—an indescribable, heavenly experience. The stories are numerous and amazing from those who have had this experience, making the spa a major attraction. http://novavita.com.ec

Nova Vita Spa and Rejuvenation Center was first established in Guayaquil in 2006, and then a new facility was built at the farm in 2011.

A chocolate masque is becoming more and more popular as the word gets out.

The essential floral water tub is a favorite attraction at NovaVita.

Essential Floral Water Spa

Only five farms in the world have essential floral water spas, and all five are Young Living farms:

- Chongon (suburb of Guayaquil), Ecuador: Palo Santo, Dorado Azul, Eucalyptus Blue, Ylang Ylang, Mastrante, Lemongrass, Vetiver

- St. Maries, Idaho: Lavender, Melissa

- Highland Flats, Naples, Idaho: Blue Spruce, Balsam Fir, Western Red Cedar, Ponderosa Pine

- Fort Nelson, British Columbia, Canada: Black Spruce, White Spruce, Canadian Balsam, Yarrow, Ledum

- Split, Croatia: Helichrysum, Rosemary, Melissa, Hyssop, Lavender, Sage

Floral waters vary depending on the plants that are being distilled at the time. Certainly, there will be other distilleries to come in the future, and many more beautiful oils will caress the skin of thousands of excited visitors.

The original school had 42 children attending for all grades with one teacher, who came two or three times a week. There was no electricity, running water, or bathroom facilities, except for a covered hole in the ground outside.

The Young Living Academy

After passing through Chongon and driving toward the jungle down a windy dirt road, Gary saw on the corner of a turn in the road, close to the farm, the "local" schoolhouse that had been built with cinder blocks in 1959. It had no water, no electricity, and only holes in the walls to let the light in, along with all the road dust; and the bathroom was a hole in the ground enclosed with broken and chipped cinder blocks. It was a deplorable sight for the 42 children who attended with one teacher for all grades, who showed up about three times a week; so Gary decided he had to change what looked to be a hopeless future for all of them.

He was told of an elderly couple that might be willing to sell their piece of ground that was half way between the school and the farm only five minutes away. The old gentleman was a bit reluctant; but when Gary offered to buy a motorcycle for him and build a small home for his wife in Chongon as part of the deal, a big grin came across his face; he happily agreed on the price; and a swampy, mosquito-infested piece of land became the beginning of Gary's vision for these poor village children.

Gary drew the plans, met with the Chongon town council, and construction began. A director was found, teachers were hired, and the doors opened in March of 2007. The school was built only for the children attending the dilapidated old school on the corner of the turn in the road just a stone's throw from the new school, but more than double that number showed up the first day. The word spread and more and more children came. Classrooms were divided and soon plans were made to expand and build onto the existing building. A year later a preschool was built, and then more additions were added for the high school.

Many students spend school time at the farm in an apprentice-type environment. Classrooms were built where they learn about Seed to Seal from the greenhouse, to soil amendment, planting, cultivating, and harvesting, as well as working in the worm houses to learn how worm casting is used to make liquid fertilizer. Other students go to the mechanic and fabrication shops, the distillery, and the laboratory. They learn about essential oil usage at the Rejuvenation Center, as

A very sad environment for children who were so eager to learn.

Gary loved playing Santa Claus for many children who had never received a Christmas present; a special moment for Gary.

Gary designed and drew the architectural plans for the school.

well as at the school; but the most popular areas of education are the distillery and the analytical research laboratory.

Some students work in the goat dairy or even in the restaurant, where goat cheese and other goat milk products are made. Extra milk goes to the school, and tons of rice are harvested at the farm for the restaurant and the school.

A large number of Young Living members have donated to the school, and the *Sponsor a Child Program* has been extremely successful. More and more children want to attend the academy that is already "bulging at the seams." Thanks to our generous members, we were able to build a high school and a covered outdoor amphitheater/gym, enabling the children to play outside underneath the gym roof while the monsoon rains are soaking the ground.

It is hard to imagine such a modern school in the rural community of Chongon, but now more than 297 children have a previously unimagined opportunity to have a bright and progressive life. In March of 2016, many Young Living members, parents, teachers, and students will celebrate the first graduating class of the Young Living Academy. The students are happy and filled with the excitement of so many possibilities for a bright and productive future.

While traveling in Ethiopia in 2010 looking for different species of frankincense, Gary visited the University in Addis Ababa and learned of a trade school that he went to see where the students were making blocks from dirt mixed with a little cement and using them to construct different buildings and student housing on the school campus. It was a very simple and inexpensive way to build; and they even made different colors of bricks, depending on the color of the soil used, which enabled them to create very interesting and colorful design patterns.

Gary was so fascinated that in his investigation, he learned that the brick-making machine was made in South Africa. He decided that would be a fabulous way to do the construction of future buildings at both the school and the farm and possibly create a different source of revenue for the farm by selling the bricks to the public.

Two machines were ordered and when they arrived at the farm in Ecuador, instructors came from South Africa to teach Gary and some of the farm workers how to operate them. It was quite successful and now thousands of bricks are made daily. It was very rewarding to be able to build the new additions to the school with these blocks for a fraction of the cost of cinder block and cement.

The Young Living Academy opened its doors in March of 2007 to almost double the number of children attending the old school. People were astounded at such a beautiful school in rural Chongon about five minutes from the farm, and now students come from all over Guayaquil, hoping to attend the Academy. Night classes for literacy are offered for the parents, who can neither read nor write, as well as beginning English.

The students learn weights and measures as they pick rosa morta flowers. This also gives them greater understanding when they go to the farm for special classes.

Gary teaches the academy students about distillation.

Construction of a covered amphitheater began in 2014 and was completed in 2015, and the high school was built in 2015. In March 2016, the Academy will celebrate its first high school graduation of 12 students.

Bricks made mostly with soil and a small amount of cement is a technique developed in South Africa that has been an educational and financial addition to the farm.

The second addition was built in 2009 for preschool, kindergarten, and a large music room.

High in the Al-Hasik Mountains, Gary made many exciting discoveries.

THE FRANKINCENSE TRAIL

Because of Gary's desire to study frankincense, he began traveling the world in 1995 to see if the stories of the famous frankincense were true. His first stop was in Oman. He was surprised to learn that *Boswellia sacra*, the resin of the sacred frankincense of Oman, had not been exported for distillation nor was it being distilled for export at that time. Frankincense resin was a local commodity used by the local people, and only a very small amount had been shipped to the royal courts of Arabia.

Gary traveled to many different countries where it was reported that frankincense trees were growing. He wanted to see and document as many different species as possible, as well as to be able to test and analyze their different compound structures. He flew to Ethiopia and for several days journeyed by air and land to the farthest villages on the borders of Ethiopia and Kenya and to the farthest northern border of Sudan.

He traveled to many unpopulated regions, where he found *Boswellia papyrifera*, which is the most commonly sold frankincense resin in the world market and makes over 75 percent of the frankincense resin sold to brokers worldwide. *B. papyrifera* is extensively used in the perfume industry and for incense. But while *Boswellia serrata*, *B. carterii*, *B. sacra*, and *B. frereana* have more than 300 scientific studies listed on PubMed, just three studies were found on *B. papyrifera*, which showed that it had minimal effect on the body.

Many companies substitute *B. papyrifera* oil for carterii oil, mixed with a little frereana and synthetic compounds to make the oil look and smell good, enabling them to offer their oil at a lower price. This is especially true of *B. frereana*, which in its pure state is triple the price of carterii.

The prized hojari resin of the *Boswellia sacra* frankincense tree. Above: Flower of the *Boswellia sacra* tree.

Early distillers are on display at the entrance of the frankincense distillery in Spain that Gary and Mary visited in 1996.

Harvesting the resin of *Boswellia frereana* takes much more work than harvesting *Boswellia carterii*. The resin of *B. frereana* is very large and can be as long as 12 to 15 inches in contrast to *B. carterii*, which is usually harvested in small pieces that are 1-3 inches long. The two different myrrh species that we offer in Young Living are thick and coarse and also take more time to harvest.

During one of their trips to Europe in 1998, Gary and Mary spent two days at the frankincense distillery in Spain. The distillation took almost 12 hours, so Gary had a lot of time to analyze and discuss the many facets of resin distillation, which is quite different from plant distillation.

Gary's mechanical aptitude made it easy for him to see and understand what was necessary for the distillation and how to build and operate the equipment. The design was in his mind, so he was ready when it came time to build in Oman.

Gary empties sacks of frankincense resin into the extraction chamber.

After the resin is put in, the chamber is filled about halfway with water because the resin needs more room for agitation in the water.

George and Gary seal the distilling chamber.

Filtering the frankincense oil.

In 1995 Gary and Mary first went to Oman to see the new excavation and archaeological findings of the ancient city of Ubar about three hours north of Salalah. This rich, opulent city was the last enjoyment and connection to civilization that the camel caravaners had before entering the Rub al Khali, or Empty Quarter. With great fear of the unknown, they began their treacherous, life-threatening journey through the vast and constantly changing desert of desolation, heat, and wind from which many never returned.

It was exhilarating for Gary because he knew that thousands of sacks of frankincense and myrrh resins had been carried and deposited, exchanged, and distilled here. Many of those camels had continued on, carrying their precious cargo to such places as Damascus, Gaza, Alexandria, Ein Gedi, China, India, and even to King Solomon in Israel.

While walking through the marketplace and seeing tons of frankincense resin, especially the beautiful and prized hojari, the highest grade of *Boswellia sacra* (Young Living's Sacred Frankincense), Gary made the statement that one day he would have a distillery in Oman and would produce this "holy anointing oil" for Young Living. It seemed impossible, but the dream was there and the vision was intense.

During the past 20 years, Gary made at least 16 trips to Oman in his desire to learn as much as he could and to verify the frankincense species of *Boswellia sacra*. He met Dr. Mahmoud Suhail, a medical doctor who also had a deep interest in frankincense. They developed a friendship in their desire to help people with this ancient oil.

After exploring so many of the Arabian countries, Gary decided to build the distillery in Salalah, Oman, the center of frankincense history, close to the harvest of the *Boswellia sacra* resin. Dr. Suhail and Gary became partners to build a distillery and started with one single distilling chamber in January of 2010; and five months later, 25 liters of Sacred Frankincense had been distilled and a second extraction chamber added. Just as Gary said in 1995, he would build a distillery in Oman, which is the first large commercial distillery for the extraction of Sacred Frankincense in modern times, perhaps since the time of Christ—a dream come true. Every year new extraction chambers have been added, now (2015) totaling seven.

The farm is located on a historical site that was once a banana plantation, with a deep-water well that provides unlimited cooling water for the distillation. The ground is perfect, and the saplings planted by our Diamonds in April 2013 are strong, healthy, and growing very well.

His Royal Majesty the Sultan of Oman sent some decorative plants and His Majesty's favorite plant, the Desert Rose (*Adenium obesum*), to be planted near the gates lining the entrance to the farm—awaiting a potential visit by His

Gary and Mary flew to Oman in 1995 to see the new excavation of the ancient city of Ubar.

Ubar was a very rich city and the last touch with civilization before the caravaners entered the Empty Quarter.

Young Living's first distillery and extraction chamber with Gary and Mahmoud in Salalah, Oman, 2010. Today, 2015, there are seven extraction chambers for frankincense (*B. sacra*), myrrh (*C. myrrha*), and sweet myrrh (*C. erythraea*).

Majesty. They also gifted the farm with five *Boswellia sacra* (Sacred Frankincense) trees and four *Commiphora gileadensis* (Balm of Gilead) trees.

The Omani government is pleased that Gary wants to help bring Oman back to its rightful place as the frankincense capital of the world—part of the original geographical area of the Hadhramaut where the *Boswellia sacra* trees grow. Young Living is currently working to establish a large plantation for growing *Boswellia sacra*.

Dr. Suhail also distilled the resin of the biblical sweet myrrh, unique to Yemen and Oman and known for its "deep woody, sweet aroma and many traditional uses for the spirit, mind, and body." Sweet myrrh, *Commiphora erythraea*, is commonly known as *opopanax*. This beautiful oil, little known to the modern world, was made available in Young Living's Exotic Oils Collection in 2014.

In 2008 Gary and Mahmoud met the Sheik of the Dhofar, who today is still responsible for the frankincense of the original Hadhramaut region in Oman.

This is the ancient tradition of cutting the frankincense tree to start the flow of the resin. This picture was taken in 2010.

Gary found this stone with an engraving of a flying serpent used by protectors of the frankincense trees to frighten resin poachers away. The stone is approximately 2,000 years old.

Along with other Diamonds, Frances planted new frankincense starts on the farm.

The largest known frankincense tree in the world, estimated to be 800 years old, in Salalah, Oman. Diamond retreat, 2013.

Gary teaching the Diamonds among the ruins of Fort Sumharam during the 2013 Diamond retreat.

Wadi Andhur, an ancient way station on the Frankincense Trail, which Gary discovered in 2009, where distilling vats are still identifiable.

Essential oils that are extracted from resins are obtained through a process called hydrodistillation. The extraction chamber is built with an agitator that spins slowly, grinding and stirring the resin in a continual motion. Approximately one fourth of the chamber is filled with the resin and then filled half way with water. The agitating action continues with about 100°C (212°F) of heat for 10 to 12 hours, melting down the resin and releasing the oil vapor into the water that is carried upward with the steam, passing into the condenser, and then flowing into the separator, which is very similar to normal steam distillation.

Today, the farm has seven operating hydrodistillers, and Dr. Suhail is currently designing a new hydrodistiller that will again increase the distillation capacity of the farm. We work with the local Shahra people to help them improve ways for harvesting and collecting the resin to secure this precious oil.

Establishing a distillery in Oman was a huge accomplishment and started Gary thinking about other possibilities in the world. This led him to explore the opportunity of partnering with others interested in distilling and being part of our production. Young Living has now built distillers in Taiwan, Japan, and Israel that are producing oils for Young Living, with several more in the viable future.

As Gary climbed the Al Hasik Mountains, he knew there had to be a way station along the way as the caravans traveled to Ubar because the distance was so far. Gary had read about Wadi Andhur, which motivated him to look for it. He was told by archaeologists at the historical museum that there was no way station, that it was just a legend, a fable. For 16 years, every time he returned to Arabia, he searched in different areas and finally in 2009 made his discovery, high on a plateau, hidden from the view of anyone passing below.

Gary took this picture of Marc and John in the brokerage house in Aden, Yemen, from where tons of resin are shipped all over the world.

Yemen—Going Into the "Forbidden Zone"

Hundreds of years ago, Queen Sheba ruled the Hadhramaut, located in modern-day Yemen, which is rich with the history of frankincense and myrrh. Gary was fascinated with Queen Sheba and this ancient civilization, which was the center of the land of frankincense through which the caravans passed.

Gary was determined to see the remains of this ancient civilization because he wanted to see and document the different frankincense and myrrh species still growing in the Hadhramaut, as well as in the surrounding countries, while traveling the ancient frankincense caravan trail from Oman through Yemen, Saudi Arabia, Jordan, and Israel.

In 2009 Gary flew to Sana'a, the capital of Yemen, to continue his research and hopefully obtain permission to travel to Shabwa, a major trade center of the ancient caravans. Unfortunately, Shabwa is in the heart of what is called the "Forbidden Zone," inaccessible to even most of the people of Yemen, let alone to any foreigner, especially an American.

While waiting, Gary explored Aden and visited many resin brokers, who were a wealth of information. He learned that only very small amounts of frankincense and myrrh resins are harvested in Yemen, and most of the resin that Yemen exports is *Boswellia papyrifera*, which is imported from Ethiopia and then shipped to other places in the world. He saw different categories and qualities of resin and how they were mixed to make the different grades very convoluted in their distinction.

Specific grades went to China and other areas of the world, but the mixed variety and lower quality was shipped to America because the broker said that Americans didn't know the difference, a sad commentary for most American brokers and buyers.

The sweet myrrh tree (*Commiphora erythraea*) on the island of Socotra.

Myrrh tree (*Commiphora myrrha*) in Yemen.

The *Boswellia elongata* tree that grows on Yemen's Socotra Island.

The minister of antiquities had been assigned to travel with Gary to help facilitate his needs while traveling in the country. Gary told the minister that he wanted to go to Shabwa. The minister was very surprised by Gary's request and told him it was impossible because Shabwa is in the middle of terrorist territory and is very dangerous. However, Gary was determined and insisted that the minister try to get permission.

Letters, conversations, and persistence continued for about two weeks, finally resulting in the permit he wanted. Dressed in native attire and with his heavy, black beard, he looked like a native of Yemen. The quiet excitement mounted as the minister picked him up at the hotel in Atac, and the journey began.

After three hours of driving, Gary entered Shabwa, the first outsider in 42 years. It was a glorious experience to see the caravan gateway through which the camels passed into Shabwa's taxation area. In those days, all caravaners had to pay 25 percent of the value of their goods to be able to continue on their journey. To avoid taxation meant death, and Queen Sheba was unmerciful.

Gary told the minister of antiquities that he wanted to go to Shabwa and was immediately told that it was impossible. However, Gary was persistent and kept sending the minister back to the government with different reasons for giving him the permit.

Gary's persistence finally paid off and about two weeks later, he received the permit to travel into the "Forbidden Zone" in the interior of Yemen with his military escort. The ancient city of Shabwa can be seen in the background to the left.

Ruins of the ancient city of Shabwa in today's "Forbidden Zone" in the heart of Yemen.

Dwellings of Shabwa's ancient inhabitants.

A 3,000-year-old incense burner from Shabwa was gifted to Gary by the Minister of Antiquities.

Standing in the ancient gateway of Shabwa leading to the taxation way station.

Ruins of the storage vaults, distillation vats, and the queen's palace were very identifiable. While digging down on the outside foundation of one of the buildings, Gary found an old resin burner, which the minister said was about 3,000 years old. As Gary handed it to the minister, the minister gifted it back to Gary, knowing of Gary's great love for the history of his country.

It was a glorious experience filled with so many unknowns and much danger, but Gary felt very blessed to be able to see this ancient city. Returning to Sana'a, he continued on his journey to the Island of Socotra, where he photographed and documented another seven species of frankincense.

His travels were a saga of many dangerous situations, unusual solutions, and determination. Some of Gary's personal experiences were so dramatic that he was compelled to write them in his journal that evolved into his historical novel, *The One Gift*, which depicts life on the camel caravans and the commanders who led them. It is an adventure of intrigue, excitement, romance, tragedy, and the portrayal of indomitable perseverance.

This novel is truly a wonderful insight into life, death, and amazing accomplishments against all odds in this historical time period. In 2104 it was listed as No. 1 on the Amazon Best Seller list. Visit: theonegiftbook.com

The talk around the fire at night was a time for stories of desert marauders and heroic tales of adventures of the caravan on the Frankincense Trail.

The Rub al Khali, or Empty Quarter, is the foreboding desert of intense sun, freezing nights, wind, endless sand, and dangers of the unknown.

Shutran, the caravan commander in *The One Gift*.

About one hundred Young Living members filmed *The One Gift* documentary with Gary.

Caravaners arrived in Petra and carried the resin sacks into the Treasury House.

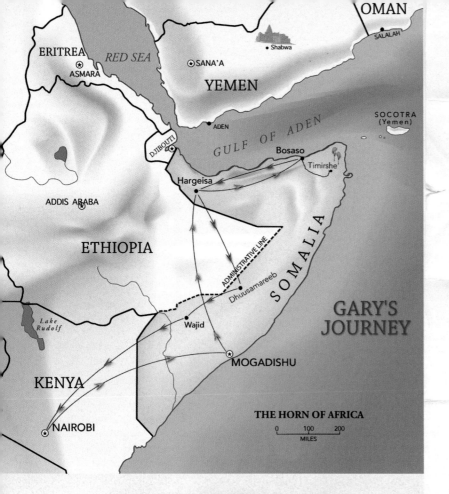

ERITREA
RED SEA
ASMARA
OMAN
SALALAH
Shabwa
SANA'A
YEMEN
ADEN
GULF OF ADEN
SOCOTRA (Yemen)
DJIBOUTI
Bosaso
Hargeisa
Timirshe
ADDIS ABABA
ETHIOPIA
ADMINISTRATIVE LINE
SOMALIA
Dhuusamareeb
Lake Rudolf
Wajid
GARY'S JOURNEY
KENYA
MOGADISHU
NAIROBI
THE HORN OF AFRICA

0   100   200
MILES

JUBBA AIRWAYS

RESERVATION CONFIRMED

0535 123910
NBO
NAIROBI
AIRLINE | FLIGHT
0535 123824
NBO
NAIROBI
AIRLINE | FLIGHT
3J | 707
3J

- RESERVATION NUMBER (PNR)   10478141
- DATE OF BOOKING   31 Oct 2013
- DATE OF ISSUE   31 Oct 2013
- PASSENGER DETAILS

Passenger Name(s)	Fare	Charges	Paid Amount	Balance
MR YOUNG DON GRAY Passport No. -	1000.00 USD	0.00 USD	1000.00 USD	0.00 USD
TOTAL IN USD	1000	0.00	1000	0.00
TOTAL IN			630.00 USD + 1358.00 AED	0.00 AED

- AGENT DETAILS
Dubai Head Office (JBW)   +971 4 2226869   reservations@jubbaairways.com
- TRAVEL SEGMENTS

FLIGHT	ORIGIN / DESTINATION	DEPARTURE / ARRIVAL	CHECK-IN FROM	CLASS	STATUS
3J708	Nairobi	Fri, 01 Nov 2013 07:00	Fri, 01 Nov 2013 04:00	Business Class C	OK
	Hargeisa -	Fri, 01 Nov 2013 11:10			
3J705	Hargeisa -	Sat, 02 Nov 2013 09:30	Sat, 02 Nov 2013 06:30	Economy Class Y	OK
	Bossaso -	Sat, 02 Nov 2013 11:00			
3J706	Bossaso -	Fri, 08 Nov 2013 08:30	Fri, 08 Nov 2013 05:30	Economy Class Y	OK
	Hargeisa -	Fri, 09 Nov 2013 10:00			
3J707	Hargeisa -	Sat, 09 Nov 2013 09:30	Sat, 09 Nov 2013 06:30	Business Class C	OK
	Nairobi	Sat, 09 Nov 2013 16:00			

- E TICKET DETAILS

Passenger Name(s)	Segment	Flight	E TICKET NUMBER
MR YOUNG DON GRAY	NBO/MGQ/HGA	3J708	5354210563036/1
	HGA/BSA	3J705	5354210563036/2

JUBBA Airways
Name: Young Don Gray
From: BSA
To: HGA
Flight:
Class:
Date: 8-NOV-13

Gate | Board Time | Seat No.

ECONOMY CLASS

الرجاء التأكد من حضورك عند بوابة الصعود إلى الطائرة قبل ٣٠ دقيقة من موعد الإقلاع

JUBBA AIRWAYS

The owner of Jubba Airways helped Gary get his visa in Somalia and flight to Hargeysa and Bosaso and then his return to Nairobi.

DAWLAD.PL
EE SOOMALIYA.

WASAARADDA AMNIGA & DDR.

PUNTLAND STATE
OF SOMALIA.

Ministry of Security & DDR.

(Office of minister)

Ref: WW/A/DDR/346/13   2/11\13

TO:-Immigration Department   BOSASO.

Sub.;- Entry Visa   Permission

The Ministry of Security and DDR Has authorized an Entry Visa for the Fallowing persons that requested to Visa in puntland Abdishakur Miree

Name	Nationality	Passport
1-Don Gary Young	USA	039704835

Therefore the Immigration offices of the airports & ports are requested to facilitate His/her requirements

This visa is permitted according to the Law And Valid for One Moth

Abdirizak Hared Ismacial
Deputy Ministry Of Security & DDR

Somalia—A Trip Into the Unknown

In August 2013 Gary had become curious about *Boswellia frereana* frankincense because of its historical prominence and all the stories told about it by another essential oil company that had been claiming for six years that frereana grew exclusively in Oman. Having spent so much time in Oman, Gary knew that *Boswellia frereana* did not and never has grown in the geographical area of Oman. The other company was unknown to any official in Oman, and its claims were upsetting to His Majesty the Sultan of Oman.

So Gary decided to distill the frereana resin to analyze its compound structure to determine possible usages. The compounds found supported its ancient aromatic use as a perfume, which has carried over into present times. The frereana resin is also softer than carterii or sacra, making it more desirable for chewing, which is very common in the Arabian countries.

Boswellia frereana essential oil is an interesting comparison to the oils of *Boswellia carterii* (Frankincense) from Somalia and *Boswellia sacra* (Sacred Frankincense) from Oman. Now that he was going to put Frereana Frankincense essential oil into the Young Living inventory, he was more determined to see the trees for himself.

There was so much mystery associated with this oil, and the information seemed to come from one person writing about what another person wrote. Gary wanted to see and know the truth for himself. He had been trying for eight years to go to Somalia and had flown to Nairobi several times expecting to go see the trees, but there was always a roadblock—too dangerous, terrorist attacks, uprising, state of anarchy, chaos, turmoil, kidnapping risk, too difficult to get to the trees, or just too far away—and so the stories came and went.

In October of 2013, while giving a Young Living presentation in Dallas, Texas, Gary met a former U.S. Special Forces military officer, who said he personally knew the president of the largest clan in Somalia, who he knew would help, opening up a new opportunity for Gary to make that journey.

A week later, Gary flew to Nairobi, where he was met by the clan president, who was currently living in Nairobi. He helped Gary pass quickly through Nairobi immigrations and customs, where an unknown future began to manifest itself.

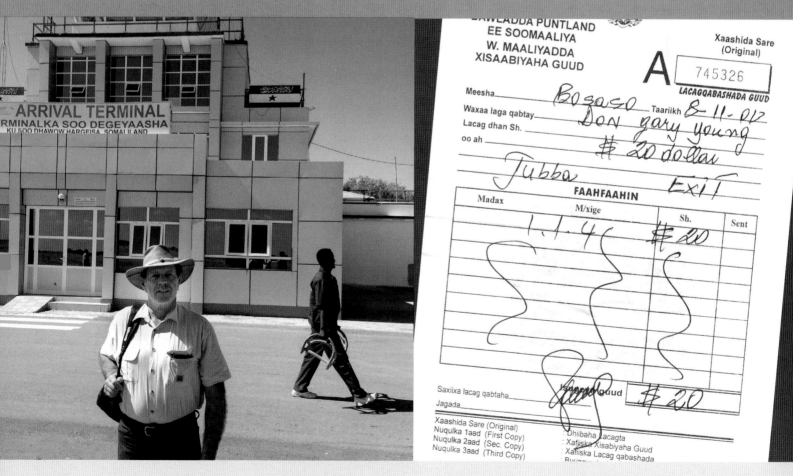

"The clan president was very helpful and answered many of my questions as he drove me to the hotel. The next morning he took me to meet the president of Jubba Airways, who was born in Somalia and said he could help me with my travels in his country. When I told them what I wanted to do, they said that it might take a few days to get the visa for Somalia; but I said that didn't matter to me if, indeed, I really would be able to fly there. I had made it this far, so I was not going to turn back and would wait as long as it took.

"But to my surprise, they called that evening and said they would be able to get my visa the next day. The night seemed long; but when morning came, we drove to the immigration office. With my passport stamped, the president drove me to the airport and said that there would be someone waiting for me when I arrived.

"When I boarded, I felt a mixture of excitement and anxiety, which seemed to stifle my breathing for a moment; but with a little self-talk, I was able to calm down; and an hour and a half later, we landed in Mogadishu, the capital.

"I sat quietly on the plane while the cleaning crew prepared for the next flight to Hargeysa (also spelled Hargeisa). Some minutes later an airport security officer boarded the plane; and when he saw me, he stared at me with a very angry look. He made a call on his radio and speaking in English said there was one American on the flight. I was uneasy and felt the hair on the back of my neck stand up, realizing that I had no place to which I could retreat for a defensive position. I felt very vulnerable and started asking myself what I was doing here.

"I turned my head to look out the window, wanting to deflect his energy. As soon as he left the plane, the other passengers started boarding; and we were soon in the air again. The flight to Hargeysa was about two hours sitting in not the greatest comfort, but I was getting closer to my goal. When I got off the plane, two Jubba Airways employees were waiting for me and quickly walked me through customs and then helped me change some money so that I could go to the market to buy some frankincense resin and get back to the hotel before dark.

"They said they would be back in the morning at 8 a.m. to pick me up; however, my departure was delayed a couple of times, so the driver didn't pick me up until 1 p.m. I was sitting in the VIP lounge at the airport waiting to board when a security officer came through and asked to see my visa. He rejected the one to Hargeysa and said I had to have a separate visa for Bosaso (also Boosaaso). I didn't know I needed two different visas for the same country. Feeling dejected, I walked back to the waiting room, wondering what I should do.

"To my amazement, the same employees who had helped me get this far came to my rescue. They led me through a side door out to the plane and told me just to get on board. I was a bit shocked when I saw the plane with several bullet holes in the fuselage that had been patched with a black caulking compound that looked relatively fresh. This plane definitely belonged in a bone yard; it wasn't even good enough for a museum.

"When the flight crew lowered the stairs, they fell off and hit the ground. I wanted to laugh as I helped them reposition the stairs. As I entered the plane, I knew I had made a mistake. I really began to question my senses and wondered why I was doing this to myself. The inside of the plane was far worse in appearance than the outside.

"Was a simple grove of trees really worth the price I was paying? If I died, would this information really mean anything to anyone? How would Mary and the boys feel? I felt my stomach come up in my throat and thought I was going to lose my almost-digested breakfast all over the plane. I took a deep breath to calm myself.

Notice the patched bullet holes in the airplane Gary flew in to Bosaso.

"The pilot came and started the engines, and the plane shook like it had tremors as it taxied to the runway. As the plane took off, all my thoughts went to my home and family. I felt I was drowning in overwhelming emotion as hot tears moistened my cheeks. I began mentally talking to myself and thanking God for my protection and that of my family so far away. Gradually, a feeling of peace and calming came over me.

"As we started the descent into the emptiness of the desert of the remote port city of Bosaso, I again began to question the purpose of coming to Somalia. I felt relief as the wheels touched down, skidding through the dirt and creating a huge dust storm; but I didn't care—I was on the ground. I could feel the tension and oppression in the air with soldiers everywhere. I knew there were only two flights a week out of here, which made me feel trapped like being held hostage.

"I was greeted very kindly by a gentleman who spoke perfect English, which didn't seem to fit in with the culture; but nevertheless, I was very grateful. He escorted me outside where three vans were waiting with soldiers who were there to guard and protect me. It was a strange feeling, making me wonder if something unusual was going to happen. I was taken to an enclosed compound that was locked down at night. I slept on a mattress on cold concrete equal to the cold water, but I was not one to complain. I surmised they were really just trying to take good care of me.

"He told me that I was the first foreigner in a very long time to come there and that everyone knew about the 'high profile American' who had just arrived. This gentleman was the president of a smaller clan of native people, but they were the ones who lived near the groves and harvested the resins. When I told him that I had come to see the frankincense trees and collect some resin to take back for research and identification, he quickly told me that he could get resin for me but that it was impossible to go to the groves.

"The groves were 400 kilometers inland to the mountains, the dirt roads were horrible, in some places were completely washed out, and traveling would take several hours with extreme risk. I told him that I didn't come this far to not see the trees; and if the vehicles could go there, the time and risk were not a deterrent.

"Seeing my determination, he reluctantly agreed but said we had to leave in the middle of the night so that no one would see us. For some reason, he said we would have to wait a day or so to get everything ready; so I spent the waiting time

Gary has a passion for helping children like these young girls, who loved the books that he gave them.

walking through the market buying resins and watching the women clean and separate the frankincense "tears" by size and color for quality grading.

Boswellia carterii has three grades, but *Boswellia frereana* has five grades, since it comes in much larger pieces of resin. The lady who interpreted for me told me that the women who have been doing the cleaning for many years looked healthier and happier than those in the village who just worked in the shops. It was most interesting.

"I bought pencils and notebooks for the children in one of the local schools that I had asked to visit. The school was dilapidated and had little to offer. My meager gifts seemed like such a small token, but they were so happy and grateful.

"It was pitch-black when we left in the middle of the night. It seemed eerie as we drove in silence down this very bumpy road, while the driver constantly looked for road signs along the way that were obscured and in many places completely washed out. After a miserable 10 hours, we could see the village on the horizon with the light of the dawn.

Stopping at a small village on the way to Timirshe to stretch their legs, the guards were close by and always watching for any unusual activity.

In some places, finding the road was guesswork.

It was amazing to see a *Commiphora gileadensis* (Balm of Gilead) tree growing next to the frankincense trees.

Gary talks with the president of the clan that guards the trees and harvests the resins.

"The village people greeted us with a big fanfare, welcoming the first foreigner they had ever seen. They were kind and loving and offered what little refreshment they had. I enjoyed our short visit, but I was anxious to get to the trees.

"I had been telling everyone about why I had developed my frankincense chewing gum; and here I was, in the home of the *Boswellia frereana* trees. It seemed that everyone I met, since landing in Bosaso, was chewing gum—frankincense gum. It was most fascinating and confirmed some of my feelings about the resin.

"The groves were 5 or 6 kilometers beyond the village, so I was anxious to go. We had driven up a wadi (dry riverbed) about 300 yards wide where frankincense *Boswellia carterii* trees were growing on the rocky sides. I gathered a little resin from a few trees that had been cut earlier. I was most surprised to see a *Commiphora gileadensis* tree (Balm of Gilead) growing right next to the frankincense.

The villagers of Timirshe had never seen a foreigner before, but they were very excited and receptive.

Boswellia carterii (left) grows out of the soil, while Boswellia frereana (right) grows out of the rocks. Yet both trees often grow near each other.

Boswellia carterii (center) often grows among the Commiphora shrubs from the myrrh family.

The Truth About Boswellia Frereana

"There is so much conflicting information about where the two species of frankincense trees grow, and now I was seeing it for myself. It was extremely interesting to see that carterii trees grow out of the ground, and frereana trees grow out of the rocks; and yet they grow practically side-by-side but are usually about 20 feet apart.

"I learned that frereana resin is difficult to gather and takes a long time because it usually takes 12 bark cuttings for the resin to really start to run, and then it is so voluminous that it runs down the entire length of the trunk. The resin is luminescent with white stripes that are exquisite. It takes about eight months to harvest because the resin pieces are so big, which is why pure frereana commands a much higher price than the carterii and myrrh resins. Frereana can be harvested only once a year, and in Somalia, only every other year.

"I collected several pounds of frereana and carterii resins to bring home, so I could distill and analyze them both to see the compound ratios in the pure oils, a bit of historical research. I also brought home different grades of carterii and frereana, so I could distill them and see the difference. I asked many questions and enjoyed the stories about the history and customs of those whose families had been the harvesters and guardians of the frankincense trees for hundreds of years.

"I thought about how I became fascinated with the exquisite, white hojari resin, the sacred frankincense of Oman, when I first walked through the marketplace in Salalah. After all my travels, it was curious to me that *B. sacra* grows mostly in Oman and Yeman, but no *B. frereana* grows anywhere in Southern Arabia. Naturally, only Mother Nature can explain that. It seems strange because we know that anciently, carterii and sacra grew in the area known as the Hadhramaut, which was under the control of Queen Sheba. *B. carterii* and *B. frereana* today grow in other areas that are now different countries. *B. sacra* grows mostly within the borders of Oman and Yemen, and the Omani hojari is unique and certainly difficult for others to acquire.

Gary is always climbing trees to get the best pictures.

Gary chewing *Boswellia frereana* resin.

The most prized *Boswellia sacra* trees grow in Oman and small, isolated areas of Yemen. The hojari resin was highly prized in ancient times, and some historians believe that this was the frankincense taken to the Christ Child. Its resin is delicately formed and is exquisitely white and almost transparent.

Boswellia frereana grows out of rocks.

"But now I had achieved my goal, and I could talk and write about *B. frereana* from my own experience, my own knowing. I had walked through the groves, touched the trees, felt the resin as it ran from the trees, talked with the harvesters—those who gather the resins—and collected a sufficient amount—about 70 pounds—of resin and was ready to go home. But, getting out of Somalia caused me a bit of uneasiness, wondering if going out would be as difficult as coming in had been for me.

"Beyond any expectation, the journey back was full of surprises: men with weapons who followed me in disguise in the market in Hargeysa from whom I hid, a shootout as I entered the airport that could have resulted in several drastic things, and even a battle with Kenya immigration officials who were forcing vaccinations on foreigners coming from Somalia. It was a good moneymaker for those in that position

of authority, and I believe that I was the only passenger who won that battle, even though it caused me to miss my flight and the frustration of having to spend more time in Nairobi.

"But when the wheels finally lifted off for the flight to Paris and then home, a sense of freedom came over me. My nine days in Somalia and three days getting out of Nairobi was a once-in-a-lifetime adventure, one that I didn't need a second time. My discoveries were priceless and I am forever grateful for my experience and for my safe return."

Gary was so enthusiastic about his trip, the discoveries he made, and the many possibilities of helping the people he met there that he wanted to build a frankincense distillery in Somaliland. However, the challenges were complex, many roadblocks kept surfacing, and it was hard to make progress. Cultural differences, the instability of the country, and difficulty in communicating made it a huge challenge to understand ideas and plans that were suggested.

Gary made the comment, "There is so much need everywhere I go that I would need to live a thousand lifetimes to do the work that is needed." He was, however, successful in being able to provide the money to build a school for the village children.

Through many twists and turns and unusual circumstances, Dr. Cole Woolley, in his travels, met a gentleman who had been distilling frankincense frereana for several years but did not intend to continue his operation. However, a large order from Young Living pleasantly motivated him to keep his operation going.

Shortly thereafter, an agreement was signed and a plan was made for the purchase of resin in Somaliland from another family member, which would be sent to the distillery, where both *Boswellia frereana* and *Boswellia carterii* are distilled exclusively for Young Living.

The distillery was rebuilt to meet Young Living's Seed to Seal standards; and in November 2014, a new high capacity distiller was installed in order to meet the Young Living demand. Both distillers are being utilized six days a week to distill the resin. The Young Living partner farm will complete the sourcing of *Boswellia carterii* and *Boswellia frereana* gum resins in Somaliland to secure sufficient quantities to meet Young Living's needs.

This is a three-generation family business that is a perfect match to join Young Living's mission to introduce essential oils into every household in the world.

3

Different Species of Frankincense

All three oils have different chemical profiles with similar and yet very different applications from flavoring to perfume, spiritual and physical elevation to chewing gum. They are all desirable to those who understand the value of pure frankincense. Here you can compare the GC analysis of the most prominent frankincense oils and their resins.

Boswellia sacra

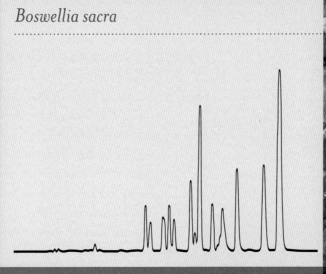

Boswellia sacra resin GC Analysis.

Boswellia carterii

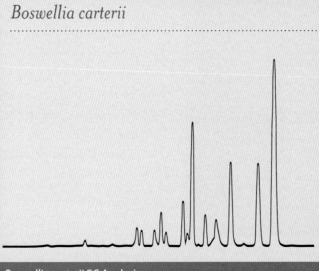

Boswellia carterii GC Analysis.

Boswellia frereana

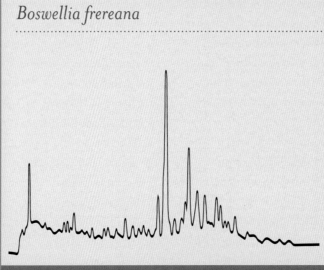

Boswellia frereana GC Analysis.

Boswellia sacra resin and tree.

Boswellia carterii resin and tree.

Boswellia frereana resin and tree.

THE HIGHLAND FLATS TREE FARM

Gary stumbled across a field of balsam fir trees in 1998 in northern Idaho in Naples, near Bonners Ferry on the Canadian border, while he was driving around looking for fields of wild tansy that he could cut and truck back to St. Maries for distillation. On the edge of a Christmas tree farm, he noticed a pile of trees that apparently was going to be burned. He stopped abruptly and ran to the field to investigate. He learned that many farmers were bulldozing down and burning the overgrown Christmas trees because the demand was diminishing every year, and the trees had grown too big to be sold, as well as the problem of spruce weevil attacking them, making it impossible to sell the trees even for nursery stock.

Gary asked if he could "take the trees in exchange for clearing their land so that it would be usable for other crops." They eagerly agreed because it was costing them between $400 and $600 per acre to have them cleared and burned. The farmer thought it strange and perhaps even a little crazy, but he was happy to let someone else take them away. It was a win-win for both.

After the first distillation of the balsam trees, there was no doubt that he had made a remarkable discovery and that he had to bring this oil to Young Living.

That began the new operation of cutting and chipping trees for distilling. It was an immense project with a complex learning curve. Determining the best way to distill to get the most oil out of the trees was challenging, but exciting. The difficulty was chipping the trees, getting the chips into the trucks, and then getting them hauled back to St. Maries. It would have been so easy to harvest the trees in the summer, but the trees didn't yield as much oil when the temperatures were warm.

Conifer trees send the oil up their branches to protect themselves from freezing in the cold—a bit like antifreeze.

When the temperatures are warm, the oil goes back into roots because it is not needed for the tree's protection. So what did Gary discover? "The colder the temperature, the better the oil."

The first harvest began in early 1998 when winter temperatures dropped to -25°F anywhere from two to five days and was very dry. Then it would start raining when it warmed up to 30°F. Out in "the bush" miles away from food and protection from the harsh conditions, cutting and chipping the trees was miserable. The chipper that worked well for tansy wasn't strong enough. Yes, for little trees, but most of the trees were big and overgrown for their purpose.

At lunchtime Gary and the crew would drive into Bonners Ferry to buy some food and then hurry to the local self-service laundromat to dry their heavy winter clothing while they ate, a blessing that probably only they could appreciate.

The first camp trailer—a welcomed lunchroom out of the cold wind and blowing snow.

The camp takes form as the walls go up.

This was the beginning of Gary's "Taj Mahal" for winter harvest, as so many members called it.

If there were any mechanical problems, someone had to drive 20 minutes to Bonners Ferry, 30 minutes to Sandpoint, 1½ hours to Coeur d'Alene, or 2½ hours to Spokane, one way. The snow and ice on the roads made traveling slow and dangerous, let alone finding the right part, getting back before dark, and trying to fix the equipment out in the field with the blinding snow and the cutting wind.

But the oil was exquisite in smell and brought an excitement that drove Gary to find its secrets. What could this oil do for mankind? In 1999 120 acres were purchased through a land auction, and a new Young Living farm was established.

A New Logging Camp

The next year a warming trailer and a larger chipper were a big improvement, but not enough. Many Young Living members came to help with the harvest, which was a tremendous help; but staying in the hotel in Sandpoint was not an efficient solution for the needs at the farm. The idea of

a logging camp began swirling in Gary's mind, and the next year a tent went up for a makeshift kitchen to serve lunch, and honey buckets were stationed nearby to meet everybody's needs, if they didn't mind baring their bottoms to a freezing-cold seat.

The floor in the dining room tent was very close to the frozen ground, making it hard to get their feet warm. During lunch, everyone hoped their gloves would dry out; however, if the gloves and coats were hung too close to the stove, they would melt. But complaining wasn't an issue. Everyone was grateful for the new improvements.

But all this work wasn't just for balsam fir. Gary found different trees in the area that he also wanted to distill, including cedar, tamarack, pine, and for the first time, blue spruce, with its amazing properties. Gary felt a lot of excitement, but it was a huge challenge with so many unexpected problems. However, he was determined and knew there was an unseen force pushing him that was greater than the problems that would come—and he was ready.

Upgrading from a regular track-hoe bucket to a grapple made loading trees into the chipper much faster and safer.

When everything goes well, it takes about 45 minutes to an hour to fill a semi-trailer.

Hot meals were fabulous, while the wood stove warmed the air and everyone's hands, but unfortunately, the floor was not heated and many feet nearly froze.

Gary's "Taj Mahal" logging camp, as it became known, began to expand.

Everyone loved Marci and the wonderful meals she made on the wood burning stove.

The learning atmosphere at camp is fun and stimulating.

The essential floral water spa is a very popular place, and the cabins are quiet, comfortable, and warm.

The rural environment of the logging camp is peaceful and uplifting and free from the noise of the busy city.

The Unstoppable Mustang Man

Going back to his boyhood experiences of logging with horses, Gary decided to haul some of his horses from the Mona farm up to the tree farm to log the ravines and areas where farmers just wanted their trees thinned, which wasn't possible to do with machines. This brought even greater excitement as Gary gave members the opportunity to either drive a team skidding the logs or the cleanup wagon that carried the broken-off branches to the chipper.

As a boy growing up on the farm logging with his father, Gary learned to drive horses first, then small trucks, and then operate equipment. These experiences served him well when he was homesteading in northern British Columbia, Canada, logging and ranching for a living. Then after his accident, looking for a way to support his family, he convinced the owner of a trucking company to retrofit one of the trucks with a hand brake and clutch so that he could drive and work.

It was fascinating to all who watched him pull into the truck yard, swing his wheelchair out of the truck, maneuver himself around to unload, then swing the wheelchair back into the truck, pull himself up into the cab, and drive out for the next load. It was an amazing feat of determination. He worked beyond normal working hours for six months, saved enough money, and went into partnership with a son and father team to buy a logging truck.

After about eight months of logging, he asked them to buy him out; and with that money and the sale of his pickup truck, he had enough money to put a down payment on his first semi, a white Western Star, and became an independent contractor, hauling logs to the sawmill in Blue River and Avola, British Columbia.

By this time, he had progressed from the wheelchair to crutches and a walker, reverting only occasionally back to the wheelchair. Even though his pain was constant and often intense, he was working and making good money. Unfortunately, after only three months, IWA (International Wood Workers Association) went on strike throughout the entire country of Canada, and all the mills were shut down.

Because Gary was an American, he went to Lynden Transport in the State of Washington seeking work, and they contracted him to haul freight from Sumas, Washington, to Mile 1202 Beaver Creek on the Alaska Highway. After seven months making a trip each week, he transferred to Fairbanks, Alaska, and began hauling on the Alaska Pipeline.

Gary's logging business before going to Alaska, 1975.

On the haul road for the Alaska pipeline, 1976.

Gary stacked and hauled trailers that had been wrecked back to Fairbanks, for which he was paid.

Thompson Pass, a 2,805-foot-high mountain gap northeast of Valdez, Alaska, averages 551.5 inches of snow per year.

Unloading a crane from Gary's truck at Prudhoe Bay.

Off-highway hauling, near Stewart Lake in Northern B.C., 1972.

Taken at the Arctic Circle at Old Man Camp, October 1976, 20°F.

The hovercraft transports Gary and his truck across the Yukon River. Notice the shiny, horse, hood ornaments on the front of his truck.

Semi-trucks moved equipment in convoys.

Gary was the youngest owner-operator at age 27 to haul on the pipeline and was known by everyone as "Mustang man." He was unstoppable and unbeatable. He hauled huge equipment weighing 70 tons over ice and snow in raging snowstorms, up and down steep mountain roads with temperatures reaching -40° to -70°F, which would have stopped most drivers. Dead Horse camp in February 1976 recorded the coldest days when temperatures dropped to -90°F without any wind chill for two nights and then warming to -82°F the third night.

In Gary's own words:

"Two other drivers and I, Sid Budden from California and Frank Bromigen from Minnesota, drove in circles for three days and nights on the helicopter landing site, stopping only long enough to refuel in order to keep our trucks from freezing up. The haul road had been closed; but if we stopped, it was unlikely that the trucks would start again and would have been parked until spring.

"The first winter, 1975-76, over 1,500 semis were left on the slopes from wrecking and freezing up, several drivers died, and many returned to the lower 48 states, defeated and broke because they lost their trucks and trailers. It was one of the coldest winters ever recorded.

"When I made the first trip north, there were 52 semi-trucks dispatched to follow the 'cat train,' which was like a caravan of earth-moving scrapers, road graders, off-highway trucks pulling fuel trailers, camp trailers, cook trailers, D8 and D9 bulldozers that leveled the tundra as they moved forward, and the road patrol that was always there to call for help if needed. Out of the 52 semi-trucks dispatched from Fairbanks to Prudhoe Bay, only 13 drivers returned with their trucks.

"Alyeska Security officers brought us food from camp, so we could keep our trucks moving. When the temperature moderated to a 'balmy' -60°F, they opened the road and let us return to Fairbanks. But the temperatures dropped again, and the devastating cold caused many accidents and problems; so the union closed the haul road again from Fairbanks to the Yukon River and requested all drivers to return to Fairbanks until the severe cold front passed and the temperatures warmed up.

"Trucks were freezing up and drivers were freezing to death. I found a company driver for Bayles and Roberts Trucking Company who was frozen to death in his sleeper north of Cold Foot Camp. Another driver, whose truck froze up north of Dietrich Camp, was found 100 yards from his truck, frozen to death on the road as he tried to walk back to camp to get help 1,000 yards from where his truck quit.

If the engine stopped, it couldn't be started, and some trucks just deteriorated away in the harsh, winter conditions.

Terrible accidents were commonplace and generally there was no recovery of the truck.

"Another time, five of us were hauling equipment 1 mile across the ice bridge on the Yukon when the lead semi-truck broke through an air pocket in the ice. The driver got wet up to his waist as the freezing water rushed in his cab. Alyeska Security was on site in minutes and pulled him from the cab, wrapped him in thermal blankets and rushed him to the helicopter site 5 miles away, in 5-Mile Camp, for emergency evacuation; but he died from thermal shock on the way. I was the fifth truck behind him and saw it happen.

"Neil Armstrong, the traffic administrator, asked if I was going to return to Fairbanks after they off-loaded me because of the road closure. I told Neil that I was here to work and couldn't afford to sit in Fairbanks and wait for the cold to pass. The union had no jurisdiction north of the river, and so I hauled and shuffled equipment between the camps for the two weeks during the extreme cold closure, again with crippling temperatures falling to -70°F.

"When I started in 1975, it was reported that over 3,500 semi-trucks started on the project. Some of the semis were company owned, but the majority of the semi-trucks were driven by owner-operators who were privately contracted to haul on the pipeline. So many drivers came from all over the States hoping to make a small fortune because the pay was so high for this demanding job; but most of them had little, if any, experience in this type of winter with such extreme conditions. Unfortunately, of the 3,500 who started, there were only 3 who finished the contract in 1977 when the pipeline was completed.

"Many times I did the mechanical repairs on my truck and forged through treacherous situations, where most would stop to wait out the storm. On the famous 'Ice Cut' just south of the Franklin Bluff Camp, I was hauling a D9 sideboom pipelayer on a 4-axle drop center lowboy. As I neared the top of the hill, I spun out with chains on both axles. I had no brakes on the lowboy and was sliding backwards toward a sheer drop-off of over 100 feet. I left the truck in gear hoping to slow down the speed of the slide when the drive shaft twisted in two jack-knifing the trailer into the bank bringing the semi and trailer to a stop just before the drop-off.

"It was about -60° that night when Alyeska Security patrolling the road found me and called into camp to have two cat bulldozers hauled down to winch my rig up on top of the hill. Parts were brought in for repairs, and I was eventually on the road again; but the whole ordeal resulted in a frost-bitten forehead, throat, toes, and fingers that later caused me to lose all of my toenails, several fingernails, and the hair on the right side of my head from touching the frame while working on the truck."

Sourdough Truck Stop, 1976.

Laying pipe at Franklin Bluff, 1976.

It is hard for most of us to imagine such experiences. Gary's ability to drive a semi-truck, operate heavy equipment, repair them mechanically, and conquer the formidable forces of Mother Nature in such an extreme environment might help explain why Gary would say that hauling chips and/or equipment from Highland Flats to St. Maries was "all in a day's work." Whatever the situation, no matter how difficult, he always looked for the solution or the answer, and he always found it. It was never too cold or icy if there was a job that had to be done.

Today, his perseverance and determination are unstoppable, just like when he hauled on the pipeline. Gary is about finding solutions and making the best decision. He never loses sight of the goal or gives up the dream.

That does not mean that he is not flexible. There will always be changes. Sometimes the weather changes the time of harvest and distillation. Sometimes the political environment of a country will change. Government policies change, import and export laws change, and sometimes it just doesn't make financial sense to continue farming or conducting business activities the current way. Gary will always make changes if he feels it will improve the process to ensure the success of the project.

Only 125 Miles to the Distillery

Learning to log with his father while growing up in the mountains to the time of his accident prepared him well for his new challenge of logging conifer trees. Although he had never distilled chips before, he knew trees and it was easy for him to organize and give directions. He had to know which trees to cut down, and then how to chip, haul, and distill them, a skill that few people have. His youth taught him well, giving him an amazing advantage for a future of which he had no knowledge.

After the chips were blown into the trailer, they were hauled to St. Maries, which was a dangerous undertaking. Gary and his friend Eldon Knittle were the only two who were brave enough to pull doubles (two trailers hooked together and pulled by one semi-truck), carrying 110,000 pounds of chips. The trips would take from four hours normally to seven wintry hours, depending on the weather. The road from Plummer to the Benewah turnoff is very windy with a lot of hairpin turns, but at least it is paved. The Benewah road up to the farm distillery is a 6-mile dirt road with narrow turns and sharp drop-offs, so drivers had to be extremely cautious.

Gary had several CDL drivers who were employees or volunteers; but as the trucks started swerving on the snow and ice and it looked like the second trailer was going to pass the first one, hearts beat faster and faster, knuckles turned white gripping the wheel, while drivers were frozen to the edge of the driver's seat with their hair standing straight up on the back of their necks.

They made the trip only once and refused to drive again. Seeing life pass before them was a heart stopper. "It was too dangerous. It was crazy. It was insane to drive those heavy trucks on those scary, narrow, and treacherous roads with snow, ice, and ferocious, blowing wind." These drivers hurriedly slithered away, apologizing in an effort to hide their embarrassment. Some commented, "I've been driving truck for 30 years and just discovered I am not a truck driver when it comes to these roads and conditions."

Gary made the drive every day and sometimes twice a day so that he could keep an eye on the logging, chipping, and distillery crews. His average downtime during the harvest was 4 hours every 24 hours, including fueling, eating, and sleeping. Gary felt they were really blessed in the 16 years they had been trucking during the winter to have lost only two trailers and had just minor damage to the four trucks.

So it was left to Gary and Eldon, but there was still tremendous worry and anxiety felt by those who knew the risks of driving back and forth. However, there didn't seem to be any other way if Young Living was going to have the oils. There were many challenges going back and forth.

Putting chains on and taking them off happened several times during just one trip between the Highland Flats Tree Farm and the St. Maries distillery.

The legal weight is 110,000 pounds, but the ice and snow in the chips at times made them heavier, weighing between 55 and 65 tons. This makes it easier to understand the magnitude of the 75 ton link-belt cranes that Gary hauled on the pipeline.

Often times, Gary would work all day harvesting and loading, and then he would jump into the semi-truck, arriving in St. Maries at midnight. But that didn't mean the trip was over and everyone could go to bed. Not a chance! That was just the halfway point of the trip.

The chips had to be unloaded on the landing and covered with a tarp before they froze into one big lump. They were left only once in the trailer, creating a very miserable time for the several people who had to work two days with pitchforks, axes, and anything else to chop out the frozen chips. So after that experience, as soon as the trucks arrived, everyone hurried to get them unloaded, which took about three hours. Once in a while Gary was able to get something to eat or catch a quick hour or two of sleep. Most of the time, though, he would help unload and then jump back into the semi and head back to the tree farm so that the trailers would be there in the morning in time for everyone to start loading again. So who got any sleep? It wasn't Gary.

Putting chains on and taking them off was done more than once during the heavy winter storms. Sometimes the second trailer had to be left at the bottom of the last hill so that the

Driving at night on snow and ice was very dangerous.

The treacherous roads never stopped Gary and Eldon from hauling the chips to the distillery.

Members come from around the world to brave the cold to be part of the Seed to Seal experience.

semi could make it up to the distillery. The usual five-minute drive when the road was dry turned into one or two hours of slipping and sliding. One bitter night, the second trailer slid and turned over at the bottom of the last hill before going up to the distillery, smashing the trailer, and spilling the precious chips in the snow all over the frozen ground. That was another miserable night. When things like this happened, it was easy to ask if the oils were worth it.

Gary made a trip with a Young Living Diamond who had come to be a part of the harvest. When they came to a turn in the road, the ice on the road had frozen so solid that the truck would not steer nor stop sliding until both trailers and the semi-truck were in the ditch leaning against the trees that held them from rolling all the way over. That night was spent with excavators, loaders, and skidders winching the trailers and truck back up on their wheels and, out of the ditch; certainly an awesome and perhaps frightening experience for a member never having had exposure to winter snow and ice.

Most people would have given up, but Gary had a mission and nothing would stop him. Having hauled on the Alaska Pipeline for two years from 1975-77 when the pipeline was being built, he knew the severe difficulties, frustrations, destruction, complete loss, and even death that came with the harsh Alaskan winters. He grew up in the mountains and understood the cold and the dangers. He had a solution for everything and was a marvel to everyone working with him, watching him fix equipment, and solving the problems. For Gary, it was just part of life.

One winter evolved into another as the oil grew in demand and the logging camp expanded. More tents went up, cabins were built, electrical lines were strung, waterlines were put in, bathrooms and showers were built, and washers and dryers started churning. The kitchen was expanded, and a beautiful, antique, wood burning stove was bought for hearty cooking and delicious meals, and the enlarged dining room served a larger number of people. Even the Internet was installed to meet members' needs.

Bigger chippers were bought and another tree sheer, another skid steer, an excavator, and more equipment kept being added. The operation grew with more people coming to the harvest to be a part of Seed to Seal, as the demand for conifer oils continued to increase.

Dr. Carlson meticulously tests the oils in the new St. Maries lab built in 2012. Gary was excited to have GC reports immediately, and everyone was excited to have the new spa addition as well, with three new hot tubs.

Balsam fir chips from Highland Flats were the first to be trucked to St. Maries and distilled.

Relaxing in the Balsam Fir floral water hot tub was a favorite place after working in the cold all day.

Construction started May 2013 with a deadline to be in operation for winter harvest the first week of January.

Gary and his family spent Christmas working through the holidays to be ready when the first group of members arrived, and they weren't disappointed.

The Highland Flats Distillery

The entire Highland Flats operation was a challenge for everyone, and Gary wondered if the trees could be grown in St. Maries to be able to eliminate the horrific drive back and forth. In 2010 Gary thought he would sell the Highland Flats Tree Farm and grow the same trees in St. Maries; but, unfortunately, the soil was different and the conifers from the Canadian border wouldn't grow in St. Maries. So Gary continued trucking the 125 miles down Highway 95 through Coeur d'Alene, up the hills, around the narrow hairpin turns, and up the Benewah dirt road to the distillery that with the harsh winter conditions was sometimes a two-day trip.

As the years continued to add up, the harshness of working with equipment in cold temperatures and the stress of just getting the job done definitely took a toll on Gary's physical and emotional strength. The rising cost of fuel and the wear and tear on the trucks continued to increase the operational costs. Gary started asking himself if there was a better way, and ideas again started swirling in his mind. He looked at the terrain and what could be done. Millions of trees would be burned to ashes, destroying God's precious oils, if he left, besides the loss of oils needed for the growing demands of Young Living.

A new distillery was the only answer. Gary saw it all. The vision was decisive and in 2012 he began drawing the blueprints and laying it all out. Over a year's time, the ideas evolved until he had on paper what he wanted. This was his opportunity to build exactly what he wanted after 30 years of preparation; and in May of 2013, the plans were ready. In August, the excavation began, forms went up, concrete was poured, and the 500 hp boiler was bought and delivered.

Digging out the reservoir at the Highland Flats Tree Farm.

Gary lifted a 7-ton beam for the hoist because the roof was put on when Gary wasn't there, and no one thought about the beam until it was too late. He either had to lift the beam or take the roof off, and the latter wasn't an option for Gary.

The cookers are 9 feet in diameter and 12 feet deep.

Gary installs the 6,500-liter chamber.

Trenches were dug and water lines were laid in the ground. Walls went up, the roof was nailed on, and the steel structure was bolted together for the 12-ton hoist. The cookers, condensers, and separators were built in the fabrication shop at the Mona farm and trucked to Idaho.

In December of 2013, excitement was high with the anticipation of completion as it all came together. On January 4, 2014, Gary distilled the first batch of balsam fir chips in the newest 21,000-liter steam extraction chamber. The distillery building is 135 feet long and 52 feet wide and houses the distillery office, boiler, extraction chambers, condensers, separators, decanting/filtering, bottling and labeling room, and a laboratory for the GC (gas chromatograph) instrument, giving the ability for instant analysis of the newly distilled oil.

A training and conference room that seats 75 people classroom style is located next to the lab, and a full fitness gym that is 36 feet by 26 feet is there for those who want to get up early in the morning and go work out with Gary.

The camp has several cabins and bunkrooms with a total of 48 beds, laundry room, a full kitchen, a dining room, a lounge/entertainment room for business and computer work with Wi-Fi Internet connection. The spa with three tubs and two saunas is just the right place to relax, visit, and share stories or just relax and enjoy the energy of accomplishment that everyone feels who comes to be part of the Seed to Seal process.

The floors of the distillery are heated, and the building is completely insulated, including the 52-foot-by-85-foot truck bay. For the first time in the history of winter harvest, semi-trucks can be driven inside and the trailers unloaded in a heated room, preventing the chips from freezing and allowing the snow and ice to melt out of them before being loaded into the cookers.

The condenser to the far right has a glass covering over the end so that the oil droplets can be seen as they pass through, flowing into the separator. Gary developed a new separator design for his distillation process. He had quartz crystals cut to specific dimensions in a marble shape. Then they were put in the separators so that as the oil passes through, the crystals tumble, increasing the separation of fine molecules that are often lost. The crystals also intensify the frequency of the oils. The GC analysis has shown as much as an 18% increase in the recovery of these fine molecules.

The first members to help distill at the Highland Flats Farm were also inducted into the construction crew.

New members will never have the experience of unloading trailers in the bitter cold. Now, when the semi-truck pulls into the heated bay, the trailer self-unloads, with a walking floor trailer, so there isn't much need for pitch forks.

A block of 12 tons of distilled blue spruce chips, which produced 14 liters of oil, is hoisted from the chamber.

Pitch forks still come in handy when the chips fall from the basket.

Everyone wants a picture in the distillery, making lasting memories of their experiences in Seed to Seal.

The completed distillery and mechanic shop, 2015.

Automation—History in the Making

In January 2014, history was in the making as the newest Young Living distillery became operational, and a different kind of winter harvest began. But there was more—something spectacular—a dream. This new distillery became the first automated, large-scale capacity, computerized distilling facility ever built in the world for essential oils. There are other automated distillers for alcohol and non-essential oil commodities, but no automated essential oil distillers exist on the scale of Young Living's Highland Flats distillery, with three 21,000-liter and one 6,500-liter extraction chambers.

The computer programming allows Gary or any other operator to set the temperature in the extraction chamber as well as the temperature of the cooling water in the condenser that sends a signal from the temperature sensor to the steam valve solenoid. The solenoid controls the steam valve that regulates the steam flow, allowing consistent temperature ramping to ensure there is no homogenizing or reflux of the oil during the extraction process. This way the cooling water temperature can be maintained in the condensers and separators, facilitating greater oil recovery.

Homogenizing or reflux occurs when in the middle of the distillation process, the temperature falls so low that the steam carrying the oil upwards to the condenser starts to turn back to a liquid state, and the oil drops out and falls to the bottom of the chamber. If this happens, the oil cannot be recovered.

In the automation process, Gary also wants to include a web video to be able to control and monitor the distillery from any place in the world with his phone. When the system is operational and perfected, it will be installed in all Young Living distilleries—another fabulous leap in raising the Seed to Seal standard. It is a dream that Gary has had for many years and has now come to fruition. Everyone who visits will stand in amazement when they see this, the most advanced essential oil distillery in the world.

Gary's great love for horses and growing up with them has added more depth to the harvest experience. He takes everyone back in time by logging and skidding the trees with his teams of Percheron horses, enabling him to protect the ecosystem of the land from the damaging effects of modern machinery. The horses are as powerful as they are magnificent looking, and they effortlessly skid the huge trees out to the chipper.

The automation system is complex, but Gary loves it.

At the landing, you'll hear the deafening roar of the monstrous 350 hp chipper as it swiftly devours the trees and spits the chips into the 48-foot semi-trailer with a conveyor-type moving floor that automatically empties the trailer at the distillery. The machine can chip trees 26 inches in diameter at 1 foot per minute and can chip two, three, and four trees at a time, taking only 45 minutes to fill the trailer, and then it is off to the distillery.

The chips come from many different tree locations where the Christmas trees are overgrown and slated to be cut and burned by owners wanting to clear the land for other ventures. The going rate for clearing the land is now (2015) between $900 and $1,000 an acre. These owners are very grateful that we have come to their area because we save them a lot of time and money, and we benefit from being able to produce oils from trees that would have ended up as ash on the ground.

Computerized distilling was only in Gary's imagination until Highland Flats.

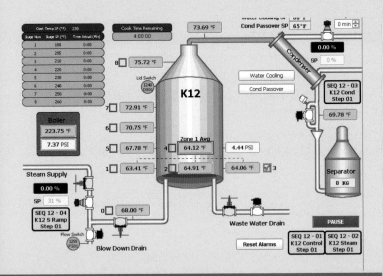

The computer interface for the automation system.

Excitement always fills the air when one of the semis full of chips arrives at the distillery. Everyone knows that means more oil that benefits hundreds of thousands of people. It is an inspiring feeling and everyone wants to help unload the chips from the trailer and then fill the distilling vats. It is fun to see how both young and old vie for the opportunity to jump into the chambers to spread and compact the plant material.

When the distilling chambers are filled to the top, everyone jumps out anxious to help close the lid. It feels like a moment of celebration when the lids are tightened and the boiler begins to crackle and pop as the steam flows into the chamber. Everyone listens to all the sounds coming from the cooker in anticipation of seeing what happens next. There is that moment of silence as everyone watching stands in awe when those first oil droplets start to bubble up into the separator followed by a burst of excitement that is hard to describe. It is truly the core of the *Seed to Seal* process, a wonderful opportunity for all members.

The steam from the 500 hp boiler softens the chip fiber, releasing the oil vapor from the fiber canals carrying them upward with the steam to the top of the chamber and into the condenser, where the steam converts back to a liquid state, the oil and water begin to separate, and the oil droplets begin to bubble to the surface of the separator. After four to five hours and throughout the distillation, depending on the material, samples are taken from each batch to be tested with the GC (gas chromatograph) instrument and the polarimeter for optical rotation to determine the quality and chemical composition of the oil.

When the distillation finishes and the oil sample is approved, the oil is poured out and taken to the decanting room to be filtered and cleaned. Then it is poured into a 500-gallon stainless steel batching tank, where it is stirred slowly for 14 to 30 days, allowing the oil molecules to harmonize and mature.

All plants have what is called a "green note," which is the chlorophyll that naturally occurs in the plants. As the paddle automatically turns, the chlorophyll gradually flashes off, leaving the beautiful, mature aroma of the oil. From there it goes to be bottled and labeled.

Gary knew the demand for the conifer oils would keep growing; so shortly after the distillery went into operation, two more distilling chambers were added so that different

Components	IBSF1312-4 15 min	IBSF1312-4 30 min	IBSF1312-4 1 hour	IBSF1312-4 2 hour	IBSF1312-4 3 hour	IBSF1312-4 4 hour	IBS #25 hr
Santene	0.16	0.13	0.18	0.18	0.16	0.24	2.00
Tricyclene	1.22	1.05	1.48	1.09	0.98	1.	1.40
Alpha-Thujene					0.15		
Alpha-Pinene	27.27	28.11	21.07	27.45	20.03	18.82	21.05
Camphene	7.52	7.03	7.28	7.78	7.06	7.56	7.48
Sabinene	1.88	1.46	1.67	1.14	1.06	1.27	1.47
Beta-Pinene	9.92	9.45	8.84	8.79	7.72	7.75	9.69
Myrcene	5.44	4.60	5.66	4.64	4.85	5.62	4.45
Delta-3-Carene	6.93	6.57	6.75	6.23	5.96	6.17	7.44
Limonene	22.03	23.95	20.87	24.96	22.61	20.72	20.53
Gamma-Terpinene	0.38	0.34	0.55	0.39	0.49	0.60	5.17
Terpinolene	1.81	1.85	2.40	1.68	2.15	2.36	2.46
Camphor	4.92	5.00	5.52	3.88	4.75	4.46	2.60
Exo-Methyl-Camphenilol	1.39	1.48	1.91	1.26	1.58	1.58	1.11
Borneol	1.19	1.34	1.88	1.23	1.62	1.62	1.17
Terpinen-4-ol	0.45	0.49	0.83	0.47	0.71	0.77	0.53
Alpha-Terpineol	0.47	0.48	1.12	0.52	0.94	1.09	0.64
Citronnelol	0.08	0,06	0.48	0.06	0.39	0.52	0.23
Bornyl Acetate	4.27	4.92	6.72	6.17	9.51	8.30	8.40
Beta-Caryophyllene	0.02	0.02	0.04	0.02	0.05	0.12	0.05
Delta-Cadinene	0.07	0.08	0.17	0.12	0.31	0.28	0.22
Tau-Cadinol	0.02	0.004	0.04	0.03	0.14	0.05	0.13
Alpha-Cadinol						0.10	
Tau-Muurolol	0.01	0.01	0.02	0.02	0.11	0.12	0.10
Crembrene	0.57	0.27	0.62	0.40	1.86	1.85	1.85
Crembrene (different isomer)	0.17	0.06	0.19	0.09	0.57	0.65	0.63
Manoyl Oxide	0.02	trace	0.04	trace	0.10	0.10	0.13
Phylloclanolide	0.03	trace	0.07	0.04	0.20	0.20	0.23
Rimuene	0.02	trace	0.03	0.02	0.10	0.10	0.11
Crenbrenol	0.11	0.05	0.25	0.13	0.83	0.81	0.87

Everyone was so surprised when the Blue Spruce oil bubbled up like pink champagne.

cookers could be dedicated to particular trees and therefore did not have to be cleaned after every cook.

Every type of tree has a different distilling time. For example, it takes 3.5 hours for balsam fir, 4.5 hours for blue spruce, 3 hours for pine, and 3 hours for cedar. The more variables, the more experimenting and testing are needed to determine the best temperature and distilling time to produce the best oil quality and yield.

It takes 12 tons of chips to produce 6 to 12 gallons of oil per cook, depending on the ages of the trees. One 32-foot semi-trailer can fill 2.5 extraction chambers of raw material, so it takes two semi-trailer loads of raw material to produce 16 gallons of oil, and that always varies depending on winter temperatures and ages of the trees.

Throughout winter harvest as temperatures changed and information was gathered with each distillation, it became evident that the colder the temperature, the greater the oil production. Each year of harvest, new discoveries are made and more knowledge is gained, making the art of conifer distillation very exact.

The hydrosol or floral water, a byproduct of the distillation, is pumped from the distillery to the spa holding tanks for the hot tubs; so people may enjoy a relaxing time soothing their aches and pains from a hard day at work. These "miracle waters," as some call them, are a favorite attraction. Two infrared saunas are also available for those desiring to "sweat out those toxins."

The entire Seed to Seal process is an amazing experience for those who come to spend time at one of the farms. It is fascinating to see how tiny drops of oil are extracted from overgrown and unwanted trees, which otherwise would be burned to the ground, make their end journey into little, brown glass bottles.

From the time our plant material enters the cooker and the oil is extracted and poured into little brown bottles, the oil never comes into contact with anything other than stainless steel or glass. This is unlike small producers in developing countries who use carbon steel, copper, and aluminum, which leave a heavy metal residue in the oils that will alter the ketones, thus changing the oil.

Learning how to log with the horses was a new and rewarding experience for everyone as they went back in time.

The Young Living logging camp is fabulous for members who have never experienced the winter harvest and Seed to Seal distillation.

Gary rides the log while skidding. It looks easy until you try it.

Winter harvest is a wonderful time for Gary to teach the boys.

Brett, the distillery manager, carefully watches the oil production.

It's always exciting when the oil droplets begin to bubble up to the top.

It is fascinating to watch the GC run the oil analysis immediately after distillation.

The filtering, bottling, and labeling room. Oils that are not bottled are poured into stainless steel containers to be transported to Utah.

Member Don Schuler fulfilled his bucket list wish to drive a team of horses. This opportunity is available to anyone who comes to winter harvest.

From around the world, members come to winter harvest to take part in the Seed to Seal process.

Highland Flats Reforestation

Reforestation has always been important to Young Living. Wherever we harvest in the world, we look for the opportunity to replenish. Planting and nourishing our ecosystem is critical to Young Living. It is so rewarding to be part of the winter harvest and then come back to be part of the spring planting. Thousands of new little tree starts can be seen at our farms in Idaho, Ecuador, and our small farm in Oman. Planting our first frankincense cuttings was very exciting, and they are doing very well. Each year as the trees grow taller, a sense of satisfaction permeates the air.

Our reforestation program is fabulous for everyone and ensures that for years in the future, we will have trees to distill for oil. The annual planting average is anywhere from 65,000 to 85,000 trees. This cycle of life is a joy to be a part of in a productive and blessed way.

Mark snapped this photo while flying the oils to Utah.

Gary, Josef, and Jacob are amazed at the growth of the 2-year-old balsam fir trees.

4-year-old Douglas fir trees in left foreground; 1-year-old balsam fir trees in background and to the right.

Members of many different nationalities and cultures become great friends working together during spring planting in the Seed to Seal reforestation project.

Replanting and replenishing the earth is a big part of the Seed to Seal promise. In the spring of 2015, 85,000 new saplings were planted.

The illumination of the northern lights is a magnificent gift to all those who come to the Young Living Northern Lights Farm in the depth of winter.

DISTILLING UNDER THE NORTHERN LIGHTS
Fort Nelson, British Columbia

In 2012 Young Living began growing at an amazing speed, and the need for oils was increasing dramatically. Black Spruce oil was a major component in several important oil blends, but the supply produced in Quebec, Canada, during the last 20 years was declining because the driving distance from the distillery was over 12 hours one way. Acquiring the trees was difficult and the road back was not easy; so less distillation was taking place, and the supply of oil could not increase fast enough to meet the needs of the growing number of Young Living members. As Gary assessed the situation, he realized that if he wanted to have Black Spruce oil for Young Living, he would have to find the trees and build a new distillery.

However, black spruce trees grow in northern Canada, where the winters are severe and the area is far from any major industry, with little access to the machinery and materials that it would take to build a distillery. But Gary felt he had no choice, so he started looking. He flew with Mark, Young Living's pilot and director of global farms, over thousands of acres of land in northern Canada from Saskatchewan to Alaska. He finally found the perfect ranch with virgin soil that had never had chemicals on it and had only a few roaming buffalo that were like the guardians of the property.

The farm is 8 miles outside of Fort Nelson at Mile 308 on the Alcan Highway, which made the access very easy. Papers were signed, the land was bought, and Gary broke ground on July 29, 2014, with the intent to beat the onset of winter. Not only did he clear land for the distillery, but he and his crew cleared land for the new fields that were planted before the ground froze in early October.

Six bison protect the farm.

Gary discovered wild yarrow growing on the farm, but there was not enough to make it affordable to harvest. So the new planting included yarrow, German chamomile, and einkorn, which to our knowledge have never before been grown in this environment. However, with the long summer daylight hours and temperatures reaching into the 90's in early June, Gary felt it was worth it to give these crops a chance.

He felt a lot of excitement when he discovered thousands of acres of wild ledum, goldenrod, and conyza (Canadian fleabane) growing in the area around the farm, although in short supply worldwide. Another discovery was the abundance of white fir and a new species of balsam fir that he was anxious to distill, which he thought could be added to the Young Living array of conifer oils.

Could Gary have imagined, as he drove down the two-lane dirt road called the Alcan Highway while hauling on the Alaska pipeline, that 35 years later he would be driving to his own farm on that same highway?

Gary breaks ground for the Northern Lights distillery on July 29, 2014.

Gary drew the plans, purchased and/or rented equipment, excavated and dug out the footings, cleared the land, built the roads, and hired the crews. Chip Kouwe, Jim Powell, and Scott Schuler, three Young Living Diamonds, answered Gary's call for help immediately. They worked side by side with Gary and Mark to get as much done as possible before the winter wind and snow came.

Even Jacob, Josef, and Mary helped before returning home for school. Jacob is a terrific excavator operator, having been taught by his father in Ecuador; Josef is learning to operate the D6 Bulldozer; and Mom took lots of pictures and brought food and water.

There were only a few weeks before winter would arrive, so working 14 to 16 hours a day was not unusual, especially with sunrise at 3:30 a.m. and sunset at 10:30 p.m. Everyone worked fast to take advantage of the long daylight hours of the far North because winter would soon turn the daylight into a very few hours.

Twenty years ago, Gary could practically take a D8 bulldozer apart and put it back together, blindfolded—and he still can.

A culvert was put in for the road going into the farm.

A lot of construction equipment was moving at the same time. It was a race against time to beat the onslaught of winter.

It was a fabulous working team that moved a lot of dirt in a few short weeks: Dave, Jim, Scott, Mark, Chip, Jacob, Gary, and Josef worked from early morning to late at night. Mary took pictures and ran for food and water.

Gary was always on the phone coordinating all the moving parts.

Since construction began in August 2014, 556 cubic meters (727 cubic yards) of concrete have been poured. The outside dimensions of the building are 150 x 72 feet, which includes the distillery, truck bay, shop, and two-story lab.

Ben Howden, a licensed contractor and Young Living Diamond from British Columbia, moved to Fort Nelson to help Gary start building. His son, Cory Howden, with 25 years in construction and cement work, took the responsibility of construction superintendent in the race to get the distillery up and running before winter. Carol and Marnie cooked three healthy meals each day, so the men could work fast with little interruption and have the nutrition needed to sustain them, especially in the cold, harsh, winter temperatures.

The forms went up and the men started pouring concrete. For weeks it went back and forth between forms and concrete. Every step of the way brought more excitement with the anticipation of firing the boiler. But winter came first and the temperatures dropped anywhere from -27°F to -44°F in mid-November. Working in the freezing temperatures was a challenge, even to the point of having to shut down a couple of times. Several Young Living members braved the cold to come and help with the construction.

Gary and Mary went to an equipment auction, where Gary bought several large pieces of equipment for the farm. It was fun as it brought back so many memories of the early days when they were building the St. Maries and Mona farms and bought a lot of equipment at different auctions.

Between the terrible cold, government regulations, and required permits, the construction was slower than expected. But finally, the last pipes were connected; three cookers were loaded; and on March 9, 2015, the boiler was fired; and the newest distillation plant went into operation.

The future of the Northern Lights Farm in Fort Nelson brings many great advantages with plans for a visitor center, campground, pelleting plant for recycling the chips, a sawmill for cutting the wood for building materials, the spa with the floral water, and the lodge, which are all scheduled for completion in the next few years. The Northern Lights Farm will become a beautiful tourist destination and a wonderful educational experience for all travelers who visit.

When the cookers finally arrived from Utah, they were off-loaded by the side of the road, waiting to be installed in their new home.

Ben's wife, Carol, and Cory's wife, Marnie, cooked three meals a day for everyone and brought lunch out to the hungry crew.

While digging out the reservoir for the distillery water supply, wells were also drilled and filtration systems and holding tanks were installed. Pipes going everywhere were laid 10 feet deep to get below the frost level and then covered with 4 inches of insulating foam before being covered over with dirt.

Cory, construction superintendent, discusses plans with Gary.

It was a huge push and even though winter came too soon, the construction went forward in spite of the tremendous cold.

This 350 hp boiler was trucked from Virginia with a smaller 100 hp boiler for a backup.

Typical winter temperatures reached an average of -20° to -30°F and on occasion even fell to -50°F.

The first of three cookers is set in place.

Ben, Cory, Wes, Gary, and the crew were quite happy with the progress being made. With the boiler and cookers installed, there was light at the end of the tunnel.

Bitter cold temperatures made the construction very difficult.

At -30ºF, the dirt froze in the bed, causing a rollover when the dirt wouldn't come out. The extreme cold created problems that were difficult and miserable to fix.

It was a fabulous day when the walls were up, the roof was on, and the heaters were installed, which eased the misery from the snow and bitter cold.

Finally enclosed, the distillery was ready for operation in March 2015.

Cutting trees at -20°F in the snow was a challenge.

A feller buncher cuts and holds 5 to12 trees at once, and when the grippers are full, it moves the trees to where the forwarding machine can pick them up and move them to the landing site for chipping. This machine has been invaluable in the severe winter conditions of the northern Canadian winter.

Forwarding trees to the chipper.

The logs are chipped on-site and then trucked to the distillery.

Members are eager to help as Gary loads the chamber.

There are two 12,000-liter cookers, one 10,000-liter cooker, one 20,000-liter cooker, and one 3,500-liter cooker, 2015.

Tanya from Yellowknife, in Northwest Territories, and Joanne and Larry from New Brunswick traveled great distances to see the first black spruce distillation.

Black Spruce Speaks

By D. Gary Young

I am here now, here to fulfill my mission as told by the ancient people. No one would hear me; but after thousands of years, I have been released to do what I was created to do for the children of this world.

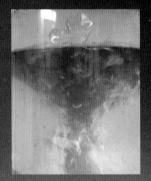

The ancient people call me 'Black Spruce of the Northern Lights' because my branches touch the dancing colors in the sky, intertwining Mother Earth's strength with the mysteries of the heavens. I bring light to the darkness of the mind, release spiritual blockages, and free the bondages that some humans call emotions, which can lead to the deterioration of life. I live in an extremely harsh environment, with challenging growing conditions, climate changes that drop to 80°F below zero, and howling winds with a wind chill of up to minus 120°F.

Summer temperatures reach 90°F, with long daylight hours that never see darkness in some areas of the northern hemisphere in dramatic contrast to the short daylight hours in the winter that don't see the sun for weeks at a time.

I live with the extremes of Mother Earth, so I can deal with the extremes of humankind. I am one of the least explored by modern man because I am looked upon as dwarfed and ugly. I am considered a nonproductive scrub tree with no value, but I can help others find their value when they are dwarfed emotionally.

Wild animals do not sleep beneath me because my habitat is so uncomfortable, and my skinny branches offer no shelter from the freezing wind and snow. Yet as they gather around me, they feel the heat I generate that increases their circulation and body temperature. I am not here to be a shelter for protection or a crutch to bear the burdens of the world. I am here to give strength that teaches self-reliance, which helps all creatures become secure, as am I.

I am here to help the human race build a relationship with the Creator, not looking to the Creator as the protector but to understand that He made all living things to have the power to be strong, adding purpose to creation, not taking energy but giving back.

Humans are co-creators with the Master of the universe and must come to understand their God-given power over all the elements, as He gave to me; to own their greatness; and to become what they were created to be. My roots can attach to anything, and I never blow over because I am so strong. My molecules bring strength to the human core, to feel grounded, and to never be defeated by the problems they encounter. Man must be able to anchor to any condition at anytime, anywhere and be able to adjust and adapt. This is what keeps me alive and will help humans find success as they partner with the Great Spirit in their journey of life.

Hold and cherish my essence. Let it help you grow, learn, and explore your God-given potential to know who you are and have the power within you to change the world with just your thoughts and to be a living example of your creation.

The Ancients treasured our sister White Spruce, which will soon come to teach another dimension of truth and knowledge. Be ready, for she works fast and is not as patient as her brother. I work more on the physical, but she works on the spiritual to help you become independent in preparing you for your partnership with the Great Spirit.

Our brother Blue Spruce is universal and grows in many different regions of the world. Our earthly powers work well together to help you find balance as you seek your highest potential in your journey through life.

The ancient people of the North look down and give thanks that we have joined together with the Great Spirit for the support and well-being of His people and Mother Earth.

The first distillation of black spruce was on March 9, 2015.

The first distillation of black spruce was a marvelous wonder as we watched the oil droplets bubble up into the separator. The action was intense as if the oil was speaking to us.

We had an awesome feeling of triumph as the oils from the first winter distillation were delivered to the warehouse. Excitement was buzzing throughout Young Living as word went out that Black Spruce oil was back in stock.

The Northern Lights distillery is a beautiful work of art, with a new state-of-the-art water-cooling and recycling system. While Gary was building the Highland Flats distillery two years earlier in northern Idaho, he was constantly thinking about his design. When the distillery went into operation, he kept watching and analyzing the flow of the distillation, which brought a new idea to him. In his mind, he could see a way to make the process better and more efficient, so he designed and built his innovative idea into the new Northern Lights operation. Once his theory and new design is proven to work the way he wants, he will start making modifications on our other distilleries around the world, which will include the rebuilding of our distillery in France.

Our Northern Lights distillery has a large, heated truck bay; so the trucks can come in and unload in the warmth, just like in Highland Flats. However, this bay is larger to accommodate the 53-foot walking floor trailers that transport the chips from the logging site. There are two 12,000-liter, one 10,000-liter, and one 20,000-liter extraction chambers, with production now running at about 530 liters each week. There is also one 3,500-liter extraction chamber for small sample distilling. From the distillery, the oils go to the filtration and decanting laboratory and then to 500-gallon batching chambers before being bottled and packaged.

The Northern Lights distillery includes a large laboratory and research center with a state-of-the-art GC instrument with dual 50- and 60-meter columns for analytical and scientific testing and documentation. Sample analysis is always completed as part of the distillation process, and sometimes multiple samples are tested during various stages of distillation to ensure that oil component percentages and other oil characteristics are optimal.

A training room that seats 100 people will be used for educational seminars and where high school and college students from around the world will come who want to study and learn distillation, organic plant chemistry, and essential oils analysis.

Gary and Dominik are excited about the growth of new helichrysum at our partner farm.

CROATIA—DISTILLING IN 19 DAYS

Croatia, the land of a thousand islands, is a paradise of aromatic plants, a beautiful land of valleys, hills, lakes, and thousands of acres of farmland on which many aromatic plants are growing. It is not only the land of helichrysum (*Helichrysum italicum*) but also of many other crops that will be distilled at the newest Young Living facility located on the outskirts of Split, Croatia.

In May 2015 Young Living purchased a two-story commercial building totaling 16,580 square meters, which includes several offices, conference rooms, warehouse space, and two large bays for the distillery. The boiler was shipped from France, the separators from Utah, and the four 4,000-liter, and one 6,000-liter chambers were built in Split. Dominik had four 1,000-liter extraction chambers that he was already using powered by a wood-burning firebox, bringing our total distilling capacity to 26,000 liters.

Everything was assembled in our new facility, and distilling began June 19, 2015, exactly 19 days from when Gary started to assemble everything and refit the building so that they would be able to distill during the short 55-day time period allowed by the government—a most amazing accomplishment.

After the crops are harvested, they are trucked to this new facility, where the Seed to Seal process continues with distilling, laboratory testing, filtering, and decanting. Then some of the oil goes to another area of the building, where it is bottled, labeled and shipped to our European distribution center. Bulk oils are shipped to the warehouse in Utah for bottling, labeling, and further distribution to other countries throughout the world.

This building houses the distillery, a large curing area for the plant material when delivered, and a large amount of storage for material waiting to be distilled. The enclosed boiler room; decanting, bottling, and labeling rooms; and a laboratory with a GC instrument for testing the oils are also on the main floor.

Management offices are on the second floor, as well as education training rooms that are available for members to use. Rooms on the lower level are for the spa with the floral water hot tubs, which will be a fabulous addition for our members and visitors, as well as rooms for massage, Raindrop, and other skin care and cosmetic applications.

This facility is planned to become a manufacturing plant, warehouse, and distribution center for the Balkans. For the first time, members and visitors alike from Europe, Russia, and the Adriatic countries can see and experience the entire Seed to Seal process on their continent and be able to participate at harvest times each year. This will surely prove to be a wonderful opportunity for thousands of our members on this side of the world.

Thousands of new starts have been germinated and are now planted.

Domesticating Helichrysum

Our partner farm is growing 78 hectares (192 acres) of helichrysum and contracting with many other farmers who have small acreages of helichrysum. It is a short drive from the farm to take the harvested crop to the distillery. Young Living Adriatic d.o.o. (proper business identification like an Inc. or LLC) was registered at the end of January 2015, giving Young Living Croatia the right to obtain more farmland, a project growing more every day. Croatia is a country very rich with aromatic plants that we are planning to cultivate with other partner and cooperative farms and that will be distilled in the new facility.

It seems that whenever Gary goes into a new country to begin a farming project, the people are skeptical and even critical; but that changes when they see him driving the tractor and tilling the ground or working with the construction crews, welding and fabricating the framework for the extraction chambers. Even teaching the boiler operator how to install the 100 hp boiler, shipped from France, was no doubt amazing to everyone after having seen only small firebox boilers.

The people are astounded as they watch him work alongside the field or construction crews. The electrician put his arm around Gary's shoulder and in very broken English said, "You are the best. You do not tell us how, you show us how. This is not normal for a business owner."

Helichrysum grows the best in extremely rocky soil, so Gary bought a 305 hp tractor and stone crusher, shipped from Germany, that had tremendous power and made the ground preparation much faster.

Transplanting helichrysum seedlings from the greenhouse.

Members help with the weeding and care of the plants.

Gary walking the fields as usual.

Helichrysum, thriving in the rocks and the humid Mediterranean climate, is almost ready for harvest.

Newly fabricated cookers and separators are ready to be trucked to the distillery.

Josef helped turn the fresh helichrysum for curing to prepare it for distillation, June 2015.

Gary designed a special system to decrease the pressure from 125 pounds to 5 pounds of pressure, enabling more steam to enter the chamber without harming the distillation process, allowing for greater steam saturation of the plant material, and increasing the extraction of the oil.

Zoran, the plant manager, and Dominik, the owner of our partner farm, join with Gary in the excitement of the success of the distillery.

Many members travel from all over Europe to see the new distillery.

Members share in the excitement of the Seed to Seal process in Croatia.

The new Young Living distillery and office building in Split, Croatia.

The first distillation of helichrysum took place 19 days after the building was purchased.

The beautiful stainless steel extraction chambers and separators were manufactured in Split.

Members from around the world attended the ribbon cutting and official opening of the distillary in Split, October 6, 2015.

Croatia Reaps Bitter Harvest From Illegal Plant Trade

Guy Norton in Zagreb November17, 2014

Croatia reaps bitter harvest from illegal plant trade Guy Norton in Zagreb November 17, 2014 A group of concerned citizens on the Adriatic island Krk held a demonstration at the toll bridge connecting their island home to the Croatian mainland on November 12 to air their concerns about the increasing environmental and financial devastation being wrought by the illegal harvesting of wild plants.

Front and centre of the good-natured protest by an assorted group of olive growers, wine producers, bee keepers, sheep farmers, war veterans and eco-warriors were concerns that the Croatian Ministry of Environment and nature protection has dismally failed to combat a growing wave of illegal harvesting of wild plants on the island, which is threatening not only environmental destruction, but is also imperilling the livelihood of traditional agricultural producers on Krk.

The main bone of contention with the protesters has been the picking of the plant Helichrysum arenarium, better known by its poetic name of Immortelle. A litre of essential oil from the plant which thrives on the rocky Croatian coast and islands can command as much a €1,700, as it is in growing demand in the cosmetics industry, especially in France, which accounts for 90% of Immortelle oil exports. The rising cost of the oil has led to an explosion of interest in harvesting Immortelle and it has been claimed that experienced pickers can earn as much as HRK10, 000 (€1,250) a month — almost twice the official average wage in Croatia.

While traditionally the harvesting of the plant has been carried out under official licenses granted to companies and individuals by the government, the lure of short-term profits has attracted a growing band of illegal pickers who the protesters on Krk claim are leaving a trail of destruction behind them. Licensed harvesters are required to abide by a strict code of conduct that involves seeking the permission of landowners before cutting off the flower heads and upper stalks of the Immortelle, which are later boiled and distilled to produce the prized essential oil – around 7,000kg of fresh flowers are needed to produce a single litre.

Illegal pickers in contrast have been guilty of criminal trespass and have simply ripped up Immortelle by the roots, destroying any chance that the plants will regrow and leading to soil erosion on the environmentally sensitive Croatian archipelago. In August this year, for example, wildlife rangers on a visit to the uninhabited island of Prvic, which is a strictly protected botanical and zoological reserve with no public access, intercepted a band of pickers who had been illegally harvesting Immortelle.

Meanwhile, on inhabited islands like Krk, gangs of pickers from nearby mainland towns such as Ogulin and Karlovac have been illegally camping out in environmentally sensitive areas of the island that form part of the EU's Natura 2000 protected habitats network, leaving behind rubbish, knocking down traditional dry stone walls and frightening livestock in their search for Immortelle.

Unenforced

In response to the environmental destruction being wrought by roving bands of illegal pickers the environment ministry announced a complete ban on harvesting Immortelle on the Croatian archipelago in September.

http://www.bne.eu/content/story/croatia-reaps-bitter-harvest-illegal-plant-trade

> *...have simply ripped up Immortelle by the roots, destroying any chance that the plants will regrow and leading to soil erosion*

The Battle to Find Helichrysum

It has been a sad realization for Gary as the events surrounding helichrysum in Croatia have unfolded. Gary first purchased land in 1996 but lost it all during the war that separated the states of Yugoslavia. In 2014 Gary realized that it was imperative to establish a farm for helichrysum. He has traveled to Croatia eight times in the last few months to establish a partner farm, purchase a building for the distillery, truck the boiler from France, and ship equipment and parts from the U.S., for what has become the largest distilling operation in that area of the world.

Until 2014 Helichrysum was harvested without restriction from June to December, when the government decided that the wildcrafted helichrysum was being depleted and dying out. So the harvest was limited from June to July for 45 days and then again from October to November for another 45 days, making it very difficult for so many farmers because their little wood-fired distilleries were too small to distill any sizable crop in such a short time. Many wildcrafters are now not able to harvest and distill enough to make the money they depended on in the past.

Because it is still legal to harvest on private land, much fighting and poaching took place. Paid by unscrupulous foreign companies, poachers went to the islands and invaded private property at night to steal the helichrysum plants, just ripping them out of the ground, stuffing them in bags, and running. They cut fences, endangering herds of goats and sheep that were freed to wander out onto the highways and be killed. Even worse, some farmers were beaten and hospitalized.

This caused the Croatian government to make a new law for 2014 restricting the wildcrafting harvest to 45 days two times a year. Then in 2015, the ruling was again changed, allowing for only one harvest for 55 days from June 18 to August 15. Presently, the wildcrafters are fighting with the government to extend the harvest. Last year, one region that was wildcrafted produced 40.5 tons of helichrysum raw material. This year, the same region harvested only 5 tons.

The government decided to auction the helichrysum by the kilo in different regions, with the bidding starting at 1 Kuna per kilo. Within the week, the price had gone up to 7.6 Kunas per kilo ($1.00 = 6.69 HRK). In addition, a Croatian VAT tax of 25 percent has to be prepaid but can be redeemed at the end of the year. A Croatian Ministry of Forestry tax also has to be paid.

Some unscrupulous brokers have hired black market laborers to poach the helichrysum before the mandated harvesting time and then smuggle it over the border into nearby towns to avoid paying the Forestry and Croatian taxes so that they could offer their helichrysum for a cheaper price. In the first week of June, according to newspaper reports, more than 70 people were arrested, and the helichrysum confiscated. These were people promising early delivery of oil in July.

Due to the shortened harvest time and small distillers, there will be a shortage of oil this year. A small amount will still be available, but many companies will either go without or have to rely on a synthetic and manipulated substitute for their oil. Therefore, knowing the source of your oil will become more and more important. Gary was blessed to be in the right place at the right time doing the right thing; and with a total of 26,000 liters of distilling capacity, Young Living will be able to produce enough oil to meet our needs.

Croatia offers an exciting future for Young Living and for those who go into partnership with us. This beautiful country, rich with aromatic plants, will become a bright light in the essential oil world and for those looking for Mother Nature's pure, unadulterated gifts. Young Living looks forward to a long and prosperous relationship with the people of Croatia.

HELICHRYSUM

Botanical Name: *Helichrysum italicum*

Components	Accepted	Rejected
	1456	V2910224
	6/22/10	8/6/12
	Area%	Area%
Alpha-Pinene	22.6	60.9
Limonene	2.7	2.1
Linalool	0.7	0.1
Neryl Acetate	5.5	Not detected
Beta-Caryophyllene	5.9	1.8
Alpha-Curcumene	2.1	Not detected
Gamma-Curcumene	19.4	2.6
Beta-Selinene	4.3	1.1
Alpha-Selinene	6.0	Not detected
Italicene	3.6	0.1

Gary and Jared visited a melissa farm in Serbia.

Gary and Mark walk through a dying helichrysum field in Serbia. The farm is in a low-land region, which was too wet and had no rocks. Because of this, the helichrysum did not thrive, as can be seen in the foreground. For this reason, it would not be good to grow crops in this region.

Oils Produced by Young Living Global Farm Operations in 2015

Ecuador, Chongon	Amazonian Ylang Ylang, Basil, Eucalyptus Blue, Cardamom, Dorado Azul, Geranium, Ishpingo, Lemongrass, Mastrante, Ocotea, Palo Santo, Plectranthus Oregano, Rosa Morta, Ruta, Vetiver
Utah, Mona	Lavender, German Chamomile, Hyssop, Clary Sage, Goldenrod, Peppermint, Spearmint, Juniper, Einkorn grain
Idaho, St. Maries	Melissa, Lavender, Tansy
Idaho, Highland Flats Tree Farm, Naples	Balsam Fir, Scotch Pine, Austrian Pine, Ponderosa Pine, Lodgepole Pine, Blue Spruce, Douglas Fir, White Fir, Red Fir (Grand Fir), Western Red Cedar, Tansy
France, Simiane-la-Rotonde	Lavender, Lavandin, Clary Sage, Rosemary, Helichrysum, Einkorn grain
Oman, Salalah	Sacred Frankincense, Myrrh, Sweet Myrrh
Canada, Fort Nelson, B.C.	Black Spruce, White Spruce, Ledum, Goldenrod, Balsam, Yarrow
Croatia, Split	Helichrysum, Rosemary, Thyme, Hyssop, Melissa, and many more to come.
Israel, Kibbutz Almog	Micromeria, Balm of Gilead, Sacred Frankincense

Following the Seed to Seal commitment, many Young Living partner and co-op farms
worldwide produce pure, genuine, therapeutic-grade essential oils.

YOUNG LIVING PARTNER AND COOPERATIVE FARMS AND QUALIFIED VENDORS

Besides our own farms, Young Living partially owns, invests in, or otherwise has contractual rights to many other farms around the world in partnership with independent growers. Young Living partner and cooperative farms have become continually more important to Young Living.

There will always be some oils in the world that we cannot grow on our own farms, so Young Living has developed partnerships with farmers throughout the world. We invest in their operation and help them become more productive and efficient. We build stainless steel extraction chambers, purchase farm equipment, teach better distillation practices, and support their operation with our technology and financial assistance.

Our partners worldwide must meet very specific requirements to produce oils that Young Living will accept. All partner operations are monitored and audited, and all incoming oils are tested before they are approved and become part of Young Living's inventory.

These respected growers work exclusively with Young Living and pledge to adhere strictly to our Seed to Seal process, where we closely monitor the quality and authenticity of the botanicals and oils they provide.

All distilleries have to meet the food grade, stainless steel quality that Gary put in place years ago, which started a whole new standard in the essential oil industry. Before that, distilleries were built with black carbon steel, which certainly would have had a negative effect on the oils. Young Living willingly subsidizes the change to stainless steel if necessary.

Through our partner farms, Young Living is able to support local, independent farming while maintaining our absolute commitment to quality.

We also purchase some of our essential oils from a variety of global vendors if their suppliers abide by our stringent, demanding Seed to Seal process. We rigorously and thoroughly test the essential oils they provide and often travel to the farm where they are produced. If any oil fails to meet the Young Living standard, the oil is rejected and returned to the vendor.

Our Seed to Seal process is the foundation of our business to which we strictly adhere. Young Living farms produce thousands of liters of essential oils a year and carefully monitor all steps of the production.

Our Seed to Seal stamp guarantees that our farms and cooperative farms have met our requirements and are the standard by which we are known throughout the world.

Gary and Dr. Lee smell and sample new oils in the laboratory in Taiwan.

Taiwan—New Aromatic Plants

In 2011 Gary Young flew to the island of Taiwan to investigate the aromatic plants of this beautiful, green, semitropical country. He met retired professor Dr. Lee, an agriculturalist who owned a small distiller and had been farming all his life. He and his son, Tiger, a young engineer, had joined together to further their agricultural projects with Young Living. Young Living purchased a 500-liter extraction chamber (distiller) and a wood chipper to help improve the farm's efficiency; and by the summer of 2012, they had distilled 130 aromatic plants and trees. Gary chose 13 oils of which two, Red Hinoki and Red Lemongrass, were produced for the 2013 Young Living convention Exotic Oils Collection.

Red hinoki, locally called hong kuai, is known for its rich, woody aroma. This aromatic tree is native to the 9,000-foot elevation rain forest that caps the tall peaks at the center of the Taiwan Island. However, the trees are harvested only near sea level in the large rivers that start as heavy streams in the steep mountains. Monsoon rains cause landslides on the steep slopes of the red hinoki rain forests that eventually fill the small streams and large creeks, widening the rivers with red hinoki logs.

The fallen logs are taken to sawmills where Young Living purchases the scrap lumber that is chipped to distill for the essential oil. This essential oil is about 95 percent composed of heavy, bioactive sesquiterpenes and is known for its ability to "build self-confidence and security" prior to physical activities.

Red lemongrass has long been grown and distilled for making soaps, shampoos, and cleaning products. By 1940 farmers had planted 140,000 acres of red lemongrass, producing 50,000 kg of this essential oil each year. However, after the Japanese occupation, the need for red lemongrass was replaced with the need for more rice and fruit orchards. In 2013 less than 200 acres of red lemongrass were being grown on fewer than 20 farms.

Dr. Lee's distillery in Taidong, Taiwan.

One of the largest remaining red lemongrass plantations was found over 1,000 feet above the seashore and is owned by the native Paiwan Tribe. The tribe had nearly 100 acres of red lemongrass ready to harvest and a new 1,500-liter steam distiller in the plains below. Agreements were made and they began harvesting and distilling this new oil for Young Living. Red Lemongrass essential oil, locally called Xiang Mao, has a beautiful and stimulating aroma and is known for its ability to "enlighten the mind and spirit for learning and focus."

For Convention 2014, the Young Living Taiwan Farm produced a third essential oil, Jade Lemon. This essential oil is expressed from green "jade-colored" mature lemons that grow only in Taiwan and China. Because more Jade Lemon trees grow in China, Young Living is partnering with growers in southern China to produce greater quantities. Jade Lemon has the aroma of lemon and lime together, which is very pleasing to the mind and is exclusive to Young Living Essential Oils.

Our Taiwanese partners continue to collect wild aromatic plants, propagate them in their large greenhouses, and domesticate them on farming land for production. A new, high capacity steam distiller has recently been installed to increase production capacity at the farm.

A grove of jade lemon trees in Taiwan.

The ruins of Ein Gedi in Israel near the Dead Sea are estimated to be from between 2500 and 2000 B.C.

Ein Gedi was an ancient way station and depository where frankincense, myrrh, and balsam were distilled. Many ointments, tinctures, and skin creams were made in the apothecary.

Israel—Growing the Ancient Balm of Gilead and Extracting Liquid Gold

Since 1991 Gary has been traveling to the Middle East because of his desire to learn more about the trees that grew anciently in this region. He traveled to Jerusalem to study Biblical archaeology at the Hebrew University and spent his weekends driving throughout Israel following stories of legends told to Gary by his professors, who had become interested in Gary's essential oil research. Driving, stopping, looking, asking questions, and following clues that were given him, he finally found the ancient Ein Gedi distillery in the Judean Mountains west of the Dead Sea in 1996.

This heightened his curiosity and gave him a whole new awareness about the Balm of Gilead, frankincense, and myrrh trees that once grew in Israel until about 1921. The original trees were first brought by camel caravan to Israel as a gift by the Queen of Sheba and given to King Solomon as an offering of peace and her desire to learn from him of his great wisdom. Sadly, as history evolved, and the great era of the caravans and the desire for the resins and oils diminished, the distilleries became defunct; and no one cared for the trees, eventually leaving them to die out.

But for Gary, Ein Gedi was a great discovery. He could see the vats, the water channels, the apothecary, and the great stone door that was still in place. His mind reeled with visions of all the activities that took place in this ancient distillery; and in his mind, he could see it all in detail as if it were just yesterday that the caravans were delivering the sacks of resin.

In Gary Young's historical novel, *The One Gift*, the camel caravans seem to come to life while carrying the frankincense, myrrh, and balsam resins and oils as they traveled from the heart of the frankincense groves in Arabia to Israel, Egypt, India, and other far areas of the world. The story culminates in Ein Gedi, where the ancient distillery looks down from the hills to the Dead Sea and where the oils of the Balm of Gilead, known historically as liquid gold; frankincense; and myrrh were extracted, and various ointments, salves, and tinctures were made.

The great stone door protected the distillery.

Gary points to the water inlet for this deep tub.

This distilling vat was filled with resin and water.

In October 2012, while Gary was again visiting Israel and Jordan, his hope was to find some balsam trees perhaps still growing somewhere in the area. He was told about a gentleman who had a small farm east of Jerusalem who supposedly was growing a few balsam trees. When Gary found the farm, he was surprised to learn that several years earlier, this man, named Guy, had become interested in the aromatic plants and trees from the desert and wanted to bring back the Balm of Gilead (*Commiphora gileadensis*) tree that produced the most prized essential oil in ancient Judean history. He had a small farm and two large greenhouses close to the Dead Sea, where he was growing a variety of desert plants and working to expand his Balm of Gilead tree project.

Resin runs from the cut bark of the gileadensis tree.

Four-year-old *Commiphora gileadensis* (Balm of Gilead often referred to as balsam) is growing well at the farm.

Beautiful, two-year-old sacred frankincense trees (*Boswellia sacra*) growing at our partner farm.

Keeping meticulous records has always been very important to Gary.

Gary enjoys the aroma of the rare Balm of Gilead.

The nursery is doing very well, and many new *Commiphora gileadensis* trees are ready to be transplanted. Josef loved visiting the farm with Dad.

Gary made an agreement with Guy for the long-term support and development of his gileadensis tree farm and the future extraction of the Balm of Gilead essential oil. A plan was made to propagate 10,000 young *Commiphora gileadensis* trees over the next two years with the financial support of Young Living.

By September of 2013, 3,500 new *Commiphora gileadensis* trees had been propagated, along with more than 100 frankincense trees, which would be planted on the future plantation. By the end of the summer of 2014, 4,500 new gileadensis trees had been propagated in the second greenhouse, altogether totaling 8,000 trees.

In 2014 Guy received a 20-acre land settlement commission from the Israeli government, located about 25 minutes southeast of Jerusalem. With the plantation established, architectural plans for a visitor center were drawn. The new farm is in a prime location and can easily welcome bus tours and anywhere from 500-2,500 tourists daily who drive down from Jerusalem to visit Israel's historical and religious sites. It is expected that 99 percent of the visitors will hear about Young Living for the first time at this farm. Such numbers will make this the most visited Young Living farm in the world.

The Young Living Balm of Gilead Farm will be the first large-scale, commercial *Commiphora gileadensis* plantation in the world. The plans for the visitor center show it connected to greenhouses, where many living aromatic desert plants from around the globe will be displayed. Visitors will be able to see how the trees were farmed anciently on terraced hillsides.

Individuals will be able to visit a great many historical and religious sites in the area. Highway 90 south will take them to the Dead Sea bathing beaches; to the hills and caves overlooking Qumran, where the Dead Sea scrolls were found in the 1940s; Ein Gedi, the ancient site of the gileadensis terraced plantation and extraction fortress; and Masada, the ancient fortress built on top of a high mountain mesa in the desert, where Judean rebels secured themselves from the Roman army but eventually committed suicide rather than be Roman slaves.

Other fascinating places to visit include Ein Bokek, the ancient gileadensis plantation and extraction facility that Josephus records was gifted by Mark Antony to Queen Cleopatra of Egypt for Balm of Gilead production, and the ancient incense road that was traveled from Petra to the Gaza Port or to Eilat on the Red Sea coast. Petra, Jordan, was one

The extraction chamber for the Balm of Gilead – *Commiphora gileadensis* or *liquid gold,* as it was called anciently.

of the richest, commercial trading centers for the caravans traveling to the far corners of the ancient world.

The history of resins and distillation is the greatest in ancient Arabia. This is where it began, where the trees grew and their resins were harvested and distilled. It is here that the uses of the oils for the supreme physical and spiritual attainment that they desired were at a pinnacle and were also subsequently lost in time to the world.

The exploration, research, and dedication of D. Gary Young has brought much of this lost knowledge back to the modern world. That which we have learned today teaches us that there is so much more to discover. His vision of what history can teach us today rewards all those who come to visit the Young Living Balm of Gilead Farm—a new beginning of an ancient science.

The fruit of the *Commiphora gileadensis* tree.

Gary teaches about vetiver distillation in Madagascar.

Gary checks the ylang ylang distiller in Madagascar.

Sadqa and her partners with Mary Lou and Gary at our partner farm in Kenya.

It is interesting to see the environmental conditions of the ylang ylang trees in Madagascar (left) compared to the beautiful, strong ylang ylang trees at the farm in Ecuador. We can imagine why Gary wanted to begin growing ylang ylang trees on the Ecuador farm.

Madagascar

Nosy Be, Madagascar, is considered by some to be the origin of ylang ylang. The ylang ylang trees from Madagascar are old and grown in nutritionally deficient soil choked with grass and weeds. For this reason, in 2011 Gary started to grow ylang ylang trees on the farm in Ecuador so that he could produce all of the Ylang Ylang oil for Young Living.

Ylang Ylang Comparison

Components	Ecuador	Madagascar	Important Components
Para-Methyl Anisole	6.12	4.07	
Methyl Benzoate	6.13	2.42	←⟶ Methyl Benzoate
Linalool	18.65	5.61	←⟶ Linalool
Benzyl Acetate	21.87	5.83	
Geraniol	0.22	1.74	
Geranyl Acetate	2.53	3.98	
Alpha-Copaene	1.23	0.59	
Beta-Caryophyllene	3.53	7.50	
Alpha-Humulene	4.12	3.68	←⟶ Humulene
Germacrene-D	15.63	22.20	
Alpha-Farnesene	5.68	15.76	
Delta-Cadinene	1.47	1.32	
Tau-Cadinol	2.40	0.65	←⟶ Tau-Cadinol
Tau-Muurolol	1.39	0.84	
E,E Farnesol	1.51	1.41	←⟶ Benzyl Benzoate
Benzyl Benzoate	8.60	6.27	
E,E Franesyl Acetate	2.02	3.10	
Benzyl Salicylate	1.28	3.18	

Pickers are paid by the weight of their baskets.

Hawaii Sandalwood Cooperative Farm

Young Living and The Royal Hawaiian Sandalwood Plantation in Kailua-Kona, Hawaii, have signed a long-term, exclusive agreement with licenses and permits from the state to support long-term sustainability. Young Living now has a secure source of this treasured essential oil that has become threatened by overharvesting.

Gary is committed to the conservation and reforestation of sandalwood trees and is excited to have a partner who offers us this opportunity to help. Planning and implementing this project is something that Gary looks forward to doing with Young Living members all over the world, similar to the reforestation project at the Highland Flats Tree Farm in Idaho. The members will have a wonderful time in Hawaii, with its beautiful weather and scenery, where snow boots, coats, gloves, and hats won't be necessary.

Josef is fascinated with the huge sandalwood root.

Young Living members walk through the sandalwood plantation.

The sandalwood distiller in Hawaii is relatively small, with a 3,500-liter capacity.

New sandalwood saplings were planted by members during the 2014 *Drive To Win* event on the Big Island of Hawaii. Each sapling was planted within a blue tube to protect it from insects and animals.

Australia Cooperative Farms—
Blue Cypress and Melaleuca Alternifolia

Gary and Mary visited the blue cypress farm in 1997, establishing the oldest Young Living cooperative farm in Darwin, Australia, located in the Northern Territories on the northern coast where the Australian blue cypress trees grow that were used to build homes and businesses. Unfortunately, the government in 1974 abandoned the plantations after a terrible typhoon made it obvious that there was a greater need to build brick and mortar houses.

The idea came to Vince, the plantation owner, to distill the blue cypress trees when he saw that the trees had so much aromatic sap. In 2000 Vince petitioned the Australian government for a patent for the process of distilling blue cypress trees that was finalized in 2002.

Vince then contacted Gary. Subsequently, Young Living provided the funds for the distillery, and Vince agreed to supply all the oil Young Living needs. Young Living has worked closely with Vince for nearly 15 years, and many Young Living members have been to Darwin to see the Australian Blue Cypress Plantation wood chipping operation and steam distillation extraction of our Blue Cypress oil.

Resin of the blue cypress is beautiful.

A forest of blue cypress trees.

Gary and Mary walk through the melaleuca forest In Australia in 1997.

The melaleuca distillery on the Gold Coast.

Quebec Partner Farm

In 1997 Gary contacted Pierre and Lucie in Quebec, who had been distilling aromatic crops since 1988. Gary asked if they would distill black spruce trees and said he would help them expand their distillation facility to increase their capacity. Pierre unexpectedly passed away a year later; but for 17 years, Lucy and her children have continued to partner with Young Living, not only for Black Spruce essential oil but also for Ledum and Canadian Fleabane oils. She invites all Young Living members to visit her farm and distillery, so they can gain their own Seed to Seal experience.

Lucie's small distillery for black spruce, ledum, and conyza in Quebec.

Gary and Lucie confer in the lab.

Black spruce chips are ready to distill.

Wildcrafting

Wildcrafting is probably the way man first harvested plants and is still practiced by small farmers all over the world. Crude tools were made out of stone and wood; and eventually, simple metal machinery was crafted that man could either pull or push to cultivate and harvest. Then the machines were adapted for horses to pull; and today "the old machines" are practically forgotten as high-tech equipment plants, cultivates, harvests, and even packages all that was once done by hand.

Machines still don't harvest on the mountainsides or down steep hillsides; and so with a basket and cutting tool in hand, the men and women go on foot to begin their small harvest.

Wildcrafting generally produces a small volume, which is not usually enough for commercial use. The same plants that have been growing in the same soil for perhaps thousands of years with absolutely no soil enhancement are not usually strong enough to be harvested every year.

It is no different than if you plant and harvest your garden vegetables every year and put nothing back into the soil. Soon the tomatoes get smaller and smaller, and the plants produce less and less. Some people choose to use synthetic fertilizer to replenish the garden soil, and some use cow manure and compost. For the first four or five years, there isn't a lot of difference. However, vegetables sprayed with synthetic fertilizers are less tasty, and the nutritional value will be different; but to the eye, they will look pretty much the same.

Many of our oils today are produced from wildcrafted plants such as ocotea, vitex, juniper, blue tansy, Canadian fleabane, tsuga, ishpingo, ledum, palo santo, myrtle, ruta, galbanum, spikenard, Idaho tansy, eucalyptus blue, copaiba, yarrow, etc. In reality, the conifer species of spruce, pine, Western red cedar, all of our frankincense and myrrh species, and Balm of Gilead could be considered as being wildcrafted. The difference is that true wildcrafting is when the plants are not cultivated for production. When we prepare the land, replenish the soil with nutrients, weed and cultivate, we are no longer wildcrafting.

Gary visited our co-op farm and distillery for blue tansy in Morocco.

The firebox is antiquated, but it is carbon-neutral, burns wood and brush found on the ground, and works very well.

Members wildcrafted lavender in France.

Wildcrafted ruta is unloaded and weighed at the Ecuador farm.

Yarrow (small, white flowers) grows abundantly all over the Northern Lights Farm. It is currently being cultivated to be harvested as a crop next year.

Copaiba from Brazil has become a very popular oil in Young Living.

When Gary first bought helichrysum in 1990, there were only 90 liters produced worldwide, which were used primarily in the perfume industry as a perfume fixative; but as he continued to discover more uses for the oil and talked about it in seminars, the demand began to increase.

Crops like helichrysum from Croatia and blue tansy from Morocco have been wildcrafted for centuries, and the gatherers have been paid by the volume or weight of the plants. With the world demand increasing, the volume from wildcrafting is not enough.

The gatherers know that in order to get paid more money, they have to produce more plants; so they harvest as fast as they can. Even the women and children help. It is much faster to pull the plants out of the ground, roots and all, and make more money because of the weight than spend the day walking the hills and cutting with a scythe or shears.

This shortsightedness hurts the farmers and has caused a tremendous decrease in the wild plants and could permanently wipe out a crop. The people who are wildcrafting are raping the land with no thought for future generations.

So, where are people going to get their helichrysum this year? Where are they going to get blue tansy in Morocco, since this country is experiencing similar difficulties? What will desperate buyers do who cannot get the oil they want? Will laboratories manufacture more nature identical oils?

It makes much more sense to begin cultivating these declining plants in fields where they can receive proper care and nourishment to grow and multiply. Naturally, it costs more money and takes more time, but the long-term reward is a strong, sustainable crop that produces beautiful plants for distilling every year.

Because of this situation, Gary made plans to find a partner farm that would grow helichrysum, which he achieved in 2015. As he began to work with different farmers in Croatia, Gary stipulated he would not buy oil that had been distilled from the helichrysum plants that were poached and/or pulled out by the roots. In 2014 he even went twice to Croatia during the harvest to make sure that only the tops of the plants were being harvested and that the whole plant was preserved and was not being pulled out of the ground.

Farmers in Ethiopia still cut and thresh grain the same way they have been doing for thousands of years.

THE ORIGIN OF PLANTS

Taken from Gary's *Power of Genuine* Presentation

Some essential oil companies tout that their oils come only from the country of origin where they are wild-crafted, which is why they do not have farms.

People who have no experience or knowledge might say, "That sounds sensible." However, how many people have knowledge of the origins of plants in the world? Probably not many—and does the origin matter?

Can alfalfa grow in France? Can it grow in northern British Columbia? Can it grow at 6,500 feet in central Idaho, in Croatia at sea level, in Washington at 1,200 feet, or in Utah at 5,000 feet? Can it grow in Montana, Oregon, California, Ecuador, Texas, Wyoming, and Serbia? Of course, it can grow in all of these places.

Where is the origin of alfalfa? Certainly it didn't start in North America, so does this mean the alfalfa in Utah is of lesser quality because Utah or Idaho are not its origin? How is it that alfalfa from the Mona area of Utah has the preferred higher protein content for the dairies in California? Does that mean that the alfalfa in Montana is of lesser quality, even though it commands a price per ton equivalent to Utah, Idaho, and Washington, as well as the far northern reaches of British Columbia like Fort Nelson, 100 miles south of the Northwest Territories border?

Does this mean that alfalfa grown in other countries around the world has lower quality because it did not originate from where it is growing? The same might be asked about wheat that grows in almost every country in the world. When you eat commercially grown wheat products, do you know if the wheat came from its place of origin? What a ridiculous idea.

Very few people would even know that wheat originates from the Mesopotamia Region in Eastern Europe, and who would even care? Does that mean we shouldn't eat wheat because it doesn't come from Mesopotamia? If you eat wheat from somewhere else, will you have a nutritionally deficient body? This line of reasoning is absolutely absurd, considering that the origin of wheat is not the United States, as most Americans think, especially when they see millions of acres of wheat blowing in the wind as they drive the freeways in Middle America.

What creates the nutritional value in wheat? Is it the place of origin, even though the land is old and and has become nutrient depleted over hundreds of years; or is it the soil condition, climate, where the crops are being grown, and how well the land has been taken care of with proper soil management and added nutrients?

Is the lavender from Idaho or Utah deficient because it does not come from France? How many people know that France is not the true origin of lavender but that its origin is actually ancient Persia, even though most of the world thinks it is from France? One might say, "That is just good marketing."

Growing and harvesting the ancient way in Pakistan.

Why did lavender grow so well in France until the last 15 or 20 years? Why has there been such a tremendous decline in the production of French lavender—true lavender (*Lavandula angustifolia*)? Why did so many fields of lavender die, and why did farmers turn to other crops or sell their farms? What happened to the lavender capital of the world, to the farmers, to the soil in which the plants grew so heartily at one time?

Did the French farmers not realize that all the chemical fertilizers and pesticides used over so many years would contaminate the soil and eventually weaken the immune system of the plants?

So what happened when a blight attacked the lavender, and then France had a severe drought that lasted for several years? Why did it not occur to the farmers that perhaps they needed to start irrigating. Why did the farmers not replenish and feed the soil with potassium and nitrogen or analyze the soil to see what the plants needed nutritionally or determine the effects of the chemicals? All this factors in to the dying lavender industry in France.

How is it that the lavender in Utah produces a higher percentage overall of linalyl acetate and a higher yield per acre? It is all about soil amendment that you cannot get in nature. Weak soil produces genetically weak plants that produce genetically poor oil quality.

Some companies claim their oils come from plants that are wildcrafted and slander people who farm and domesticate their plants and build the soil with organic and wholesome nutrients. Healthy soil produces genetically healthy plants that produce high quality oil with well-balanced molecules for a strong immune system and reproduction. When plants are cared for and nutritionally fed, they are able to do what God intended for them to do, to keep our planet balanced and to support humankind. This ability of the plants is greatly reduced when they are genetically altered and damaged with synthetic chemicals.

Any grower wanting to produce high quality oil must know how to analyze the soil, plow it, plant seeds or seedlings, and then cultivate the crop. How would someone who has never analyzed the soil, plowed and planted it, and then cultivated newly growing aromatic plants even know if they are growing in the right conditions, let alone never having distilled an aromatic crop before going to tell the world that they are investing in cooperative farms.

In what are they investing? Crop seed? Fertilizers and pesticides? Equipment for irrigation, planting, harvesting, or even distillation? How can they say they buy their oil only from crops grown and distilled in the place of origin, yet they do not even know the real country of origin? To really understand origin, we would have to follow the evolution of planting and harvesting from people and cultures of ancient times, the same place that wildcrafting began.

Where would the world be if we depended on the wildcrafting of wheat, corn, beans, potatoes, etc.? The world would be starving.

Thirty to forty years ago, many plants were wildcrafted, even lavender. Gary learned to wildcraft lavender and thyme with Mr. Viaud and then distilled them in Mr. Viaud's distillery on top of the mountain. One of the many things that Mr. Viaud said was that wildcrafting was a thing of the past.

Essential Oil World Statistics

The world demand for essential oils is a thousand times greater today than 30 years ago, even 5 years ago; and even though more oils are being produced, there is not enough to meet the demand for pure oils.

Essential Oil Trade Statistics give us some very interesting information:

1. "Total world trade in essential oils is slightly below US $4 billion about 1% of global trade." http://www.sadctrade.org/files/Essentials%20Oils%20TIB.pdf

2. "The total world fragrance and flavour market was estimated for 2010 to be approximately USD 22 billion, a 22% growth from 2007" (according to Leffingwell & Associates). http://www.leffingwell.com. [International Trade Centre: http://www.intracen.org/itc/sectors/essential-oils/]

3. "According to the United Nations Comtrade statistics, the size of essential oil fragrance and flavor global market was estimated at US $24 billion in 2011, growing at an annual rate of 10%. The major consumers in the multi-billion dollar global essential oils market are United States (40%), Western Europe (30%) and Japan (7%), with trade in essential oils and related products increasing at about 10% per year. The United States is the largest importer (US$ 2,721 million) and consumer of essential oils, with consumption equaling about 40% of the total production." African Natural Plant Products Vol. II, Discoveries and Challenges in Chemistry, Health, and Nutrition (ASC Symposium Series), H. Rodolfo Juliani, James E. Simon, Chi-Tang Ho (Eds.) American Chemical Society, 2014, p. 289.

4. 12 March 2015 Report by Kusnandar & Co, IPR Attorneys with offices in Jakarta, Bali, Singapore, Hong Kong, and Shanghai. "The world essential oil trade value is more than USD 4 billion, with an average growth rate of about 5% per year."

The following terminology is very helpful in understanding the many methods of testing and the instruments used.

Optical Rotation is determined with a polarimeter, which characterizes or identifies the optical activity of a substance. The instrument measures the rotation of a polarized light source passing through the substance. Almost all oils contain optical active compounds that affect the rotation of light. However, if an equal mixture of chiral (mirror image or optical isomers) molecules are present within an oil (which usually does not happen in nature), the optical rotation reading will be zero.

A study on the discovery of the chiral differences in the two frankincense species carterii and sacra was published in 2012 in the *Journal of Chromatography A* by D. Gary Young and his co-authors. Sacra contains a majority of optical isomers that have a dextrorotary or (+) form, and carterii contains a majority of optical isomers that have a levorotary or (-) form. This was the first time that the two frankincense species have been shown to be different.

GC (Gas Chromatography separation) is a technique whereby a complex mixture of molecules is separated into individual molecules.

GC/MS (Gas Chromatography/Mass Spectrometry) is where mass spectrometry is coupled to the GC instrument and is the means to allow identification of the molecules that are separated by the GC.

GC/IRMS (Gas Chromatography, Isotope Ratio, and Mass Spectrometry). When added to the GC/MS analysis, this additional step involves placing the material to be identified into a combustion chamber where under high heat (1500°C/2732°F), it will completely break down into the components carbon, hydrogen, oxygen, and nitrogen (if present). These are the basic building blocks of matter. Young Living purchased standards that will identify the *ratios of isotopes* and determine if those ratios are natural (created by plant metabolism) or synthetic (created in a lab). In other words, the IRMS allows us to identify whether a sample is natural or synthetic.

HPLC (High Pressure Liquid Chromatography) is used to separate, identify, and quantify the individual components in a material. Unlike the GC, the HPLC analyzes liquid mixtures that contain compounds that are difficult to volatilize or cannot be evaporated out of the mixture. The HPLC is ideal for separating large molecules like vitamins, hormones, and other biomolecules like synthetics.

The Refractive Index measures the penetration of a certain wavelength of light through a medium. The measurement varies from one material to another and is used to identify different substances.

The term pH is used to measure the activity of hydrogen ions in solution. It is used to determine if a substance is acidic, neutral, or alkaline. The pH scale ranges from 0-14, with 7 being neutral.

A Brix instrument measures the sugar content of an aqueous solution. A high degree of Brix within a plant indicates that a high oil content is present and helps to determine the best time for distillation.

FTIR or NIR (Fourier Transform Infrared Spectroscopy/Near Infrared) is used to help identify oils, dietary supplements, and personal care products by measuring and graphing the absorption of light across a wide array of wavelengths. The resulting graphs are then compared to a known library for identification.

Specific Gravity is the ratio of the density of a substance to the density of water.

Viscosity describes the thickness of a liquid substance. A thick substance has a higher viscosity than a thin or runny liquid.

Microbiological tests are used to identify and test for any pathogens or undesirable microorganism that may be present in products. These tests ensure the safety and quality of that product. The microorganisms we test for are *Escherichia coli, Staphylococcus aureus, Pseudomonas aeruginosa, salmonella,* and coliform, along with yeast and mold.

Combustibility measures the flash point of an essential oil. The flash point is the lowest temperature at which there will be enough flammable vapor to ignite when an ignition source is applied. Once the company determines the flash points, it must list the flash points that are problematic in their Material Safety Data Sheets (MSDS) for safe shipping.

Mr. Viaud said you can't harvest enough by hand to make any money, and the time would come when wildcrafting would stop. He was certain that the governments would stop it because they would see the natural resources being depleted. He couldn't stress that more when in 1992, he said, "Mr. Gary, the time will come that if you do not grow it yourself, you will not have pure oils."

Carbon Dioxide (CO₂) Extraction

CO_2 extraction, also known as hypercritical carbon dioxide extraction, is becoming a popular method of production because the plants are distilled at a lower temperature with ranges between 105-140°F. The plant material is packed in the cooker; and then using high pressure, the CO_2 gas is released into the chamber to saturate the plant material, causing it to release the oil. When pressure returns to normal, the CO_2 gas dissipates.

This method is preferable over using chemical solvents and could be useful for extracting heavier compounds; however, CO_2 extraction is done only in limited amounts, certainly not in the volume of steam distillation. However, Gary is not in favor of using the CO_2 extraction method, and it certainly cannot work for the volume Young Living needs.

Adulteration and Chemical Manipulation

More people are becoming educated about synthetics, so laboratories are accommodating them with what they call "nature identical oils," and no one asks what that means. Unfortunately, it sounds good but it is just another synthetic oil.

A common practice by brokers is to buy third- or fourth-grade oils and then cut them with synthetics and sweet chemical fragrances. Pure essential oils have distinct, individual aromas; and they are not all sweet. When an essential oil smells sweet like candy, you can suspect that it has been adulterated to make its smell more appealing. Pure oils do not smell like candy but have their own individual aroma.

Will a little synthetic hurt you? That depends on your immune and elimination systems, but why take that risk? Synthetics accumulate in the liver, reproductive organs, and fat cells; and some have a shelf life of more than 100 years.

The other way essential oils are adulterated for profit is to add inexpensive oils that have similar constituents. For instance, cinnamon bark oil can be adulterated with cheaper cinnamon leaf oil. This can be detected by analysis because leaf oil contains a higher content of eugenol, but how many companies take time to check for purity?

Cheaper cornmint is used to dilute peppermint oil, but peppermint's menthofuran content should be from 0.4 to 14.6 percent, while in cornmint it is not detected or detected only in levels up to 0.01 percent. The biomarker viridiflorol is found in peppermint up to 0.9 percent, while it is not detected in cornmint at all. Those who spend the time and money to analyze oils will easily spot such adulteration.

Be sure you understand that the scientific instruments needed to detect adulteration require a significant investment. For instance, university scientists from Italy used GC/C/IRMS in combination with GC/MS and GC/FID analysis on 19 commercial samples of rose (*Rosa damascena*) oil and found unusual delta (13)C values in most of the oils, indicating that the much cheaper palmarosa oil (*Cymbopogon martinii*) had been added. But how many essential oil companies invest in these scientific instruments to ensure purity?

The essential oil industry has become a big "money maker" because the unknowing public has no way of determining the quality or purity of the oils they buy. Most people go by the smell; and if it smells nice, "It must be good." Besides that, who would know how to have an oil tested? Where would you send it and what would it cost? The average person would not generally entertain these questions.

Today, it is not unusual to find a variety of essential oils listed as part of the ingredients on the labels of hundreds of consumer products from skin care and cosmetics, to household cleaning products, and to a vast array of food and animal products. But how does the consumer know the quality of the oil listed on the label? Naturally, there isn't an easy way to know. But for the volume of oil that is used in the millions of products that are on the market, it would be difficult to say that there are enough oils produced in the world to meet the demand.

Young Living invests in the farmers, buys land, invests in equipment, plants crops, and builds distilleries to not only secure our oil production but to also support local economies and create security for rural communities.

With this in mind, one has to be prepared for certain oils to go out of stock until the next harvest and distillation. It seems that we all want it when we want it, but the choice is to either wait until the next distillation to have a pure oil or buy a not-so-pure or synthetic oil.

Gary Young has spent 30 years developing relationships with people all over the world. He has often said, "My quest has always been to develop a strong foundation for the future supply of our members and their children for generations to come." For this reason Young Living will always be the world leader in essential oils as new farms are established to provide for the essential oil demand of our members. The strength of our dedication, sacrifice, and determination is the foundation of Young Living and our commitment to you.

Scientific Analysis

Using scientific instruments, analytical chemists are able to know an oil's compound structure and determine how they might be used for aromatic, physical, emotional, and spiritual benefits, as well as in the vast flavor and fragrance industries. They are able to test for a "balanced" chemical profile and also detect additives that are either natural or synthetic and/or the manipulation of the molecular profile of a pure oil.

When the chemical profile is different from the accepted library profile, the chemist knows that something "extra" has been added to the oil. A good chemist will often be able to tell what substance may have been added that will increase or decrease certain percentages of a compound.

Over the years, Young Living has added an essential oil library containing over one million index files for analytical comparisons. This library includes data purchased from CRNS in France that Dr. Hervé Casabianca has worked on for decades and the compounds Gary Young has identified in his research.

We purchase "libraries" with our GC/MS instruments. For Ecuador alone, we bought a library of 500,000 profiles. When the GC/MS analysis is completed, we compare it to our database and also that of Dr. Casabianca. Then we check our retention indices library, which is a type of "address" that shows where a compound "elutes" or comes out on the GC analysis. It is a great backup to firmly document a compound.

The experienced analytical chemists at Young Living are amazing, and their experience combined adds up to about 185 years of experience on these instruments.

Synthetic components are also detectable through such instruments such as IRMS, optical rotation, and HPLC. Young Living uses many test modalities to obtain analytical scientific data. Oils coming from our farms we know are pure; but with the GC/MS testing, we can see in the constituent profile the effects of the variables of the rain and sunshine, the correct time of harvest and curing, the time and temperature of distillation, as well as the nutritional content of the soil in which the plants and trees are growing. This information is extremely valuable in being able to determine the exact time and process of distillation to obtain the highest quality possible.

When an oil is "created" through the combination of different oils or individual compounds, the analytical profile

Dear sir,
I'm a French exporter who exports products from Provence (south of France) to North America.
select very typical products which quality is unquestionable. would like to know if you could be interested by Provencal products like lavender essential oil made in France?

(50% lavender & 50% lavandin) 100% pure.
15ml=$2
50ml=$4
100ml=$7
Bulk = $65/liter

hope to hear from you soon
Yours sincerely,

This gentleman probably gave no thought to the fact that he was selling a cut or mixed oil—definitely not a pure lavender. Cutting the lavender with lavandin is a well-known practice in the industry. Just business as usual.

will be very different from the accepted library index. If the oil is made with adulterated components, the profile will not match or be comparable to the library index. One oil can even be a combination of different species of the same genus.

For example, an oil can be sold as "frankincense" yet be a blend containing a combination of frankincense species: *Boswellia carterii*, *B. frereana*, *B. papyrifera*, *B. neglecta*, etc. Varying component molecules can be identified. An example is finding epi-lupeol and lupeol in a frankincense oil, which will tell you that frereana is included. This compound is not found in other frankincense species. Individual compounds in such a blend can be identified, but they will not resemble any single oil.

If there is any question about any oil coming from a partner farm or vendor, then more instrument testing and analyses are used to be certain of the oil purity.

The Day of the Laboratory

From ancient times moving forward into the 19th century, a laboratory-made oil was not something anyone would contemplate. However, during World War II, the warring countries were not able to ship goods out of their countries; and so the United States lost its essential oil supply from Europe. This might be one of the reasons the chemical industry began to develop fragrances and flavorings in the laboratory. They discovered that these chemical creations were easy to make, cheaper, and more consistent in their "quality." People didn't seem to mind or they simply had no awareness as laboratory flavors moved into the food industry; and fragrances moved into cosmetics, soaps, and common everyday products.

Today, chemists use highly technical methods to synthesize oil molecules to increase the volume and change the aroma. The day of the laboratory has become highly sophisticated where anything can be modified and changed, and it is almost impossible to detect any kind of compound alteration.

Adulterated or modified oils are difficult to detect on a GC/MS because these instruments cannot determine if the molecule is natural or synthetic. Highly sophisticated IRMS analysis or a "trained nose," most well-known in the perfume industry or someone like D. Gary Young, can usually determine the ratios and percentages of essential oil compounds and detect the presence of synthetics; but people like this are a very small minority.

Nature-identical oils are synthetics made entirely in the laboratory and are used mainly in cosmetics and food flavoring but are also found in the aromatherapy industry. It is truly fascinating how this industry's standards have lowered to accept synthetic chemical additives.

Adulteration can occur anywhere along the production process and is disguised in so many ways. Synthetic or adulterated oils can be mislabeled, diluted, and adulterated with various chemicals and cheaper oils to increase volume and change the aroma.

Common Adulterating Agents

Clary Sage	Cut with synthetic linalyl acetate, linalool, lavender oil, bergamot mint oil
Geranium	Palmarosa, citronella, various synthetic fractions
Lavender	Lavandin from which the camphor is extracted, synthetic linalyl acetate and linalool added
Frankincense	Frankincense (*Boswellia carterii*) can be created with many frankincense composites and synthetic fractions
Melissa	Citronella, lemongrass
Neroli	Lemon, lime, orange, petitgrain
Peppermint	Cornmint up to as much as 85%
Rose	Cut with palmarosa, citronella, and various fractions, both synthetic and natural additives
Rosemary	Camphor, eucalyptus, sage
Sandalwood	Amyris, araucaria, cedarwood, castor, copaiba, glyceryl acetate, benzyl benzoate, and other synthetic additives
Ylang Ylang	Other Cananga species, balsam, copaiba, various other fractions, and synthetics

Although the chemical industry has become very big and powerful, awareness is growing among people who don't want chemical adulteration. They want to be healthier and happier, living with greater wellness, and are looking for any and all products that are unprocessed or with less processing, wanting to live in a cleaner environment free of chemicals—and that is a tall order in today's world.

However, health food companies and grocery stores are thriving as they offer organic foods and chemical-free products to the public. Health food products are scattered on the shelves of well-known grocery stores, and even the average convenience stores are beginning to add better choices to their inventories with healthier snack foods and drinks. People are more aware of the dangers of chemicals to their bodies and the rampant abundance of toxins in our environment. More people are committed to living in vibrational-clean surroundings, which has to start within our own bodies.

Essential oils are a part of nature, and our bodies love what comes from nature. Our bodies know how to respond instinctively. Whether we touch it, feel it, or breathe it, the body is refreshed, revitalized, uplifted emotionally, strengthened spiritually, and feels greater life awareness with the life-giving pure essence of the plant kingdom. It is like water. We cook with it, we bathe in it, and we drink it without asking questions. It is a natural part of our lives.

Essential oils are God's gift to mankind, and we must respect them by keeping them pure to support our well-being freely as God intended.

Myrrh

Botanical Name: *Commiphora myrrha*
Country of Origin: Somalia

Components	Accept C0527 6/18/14 Area %	Reject 72713 7/20/14 Area %
Beta-Elemene	3.7	3.0
Curzerene	22.3	24.6
Germacrene B	1.8	2.0
Furanoeudesma-1,3-diene	38.9	22.9
Lindestrene	11.5	11.8
2-Methoxy Furanogermacrene	6.0	Not Detected
Other Notable Components		
Diethyl Phthalate*	Not Detected	4.6

Vendor sample 72713 rejected due to lack of 2-Methoxy Furanogermacrene, and other components not being with specification, as well as the presence of Diethyl Phthalate. Diethyl Phthalate is a commonly known plasticizer. Contamination may have occurred due to interactions with certain plastics or plastic adhesives during the distillation or storage process.

Franken-Yeast Creates Expensive EO Molecules

Since "nature identical" products are no longer accepted as consumer friendly, it was just a matter of time until a newer technology arrived. The newest and cheapest way to make flavors and fragrances is now "Synthetic Biology."

High value molecules (key components of essential oils) can be recreated by fermentation from sugar with Streptomyces enzymes or by creating synbio yeast (synthetic DNA is designed on a computer and inserted into the DNA of naturally occurring yeast) to produce a scent molecule like vanillin. Synthetic vanillin is added by unscrupulous vendors to thinned-down lavender and peppermint to make them smell and taste sweet like candy.

A September 2015 study stated: "With the tools of metabolic engineering, microorganisms can be modified to produce compounds such as esters, terpenoids, aldehydes, and methyl ketones."[1]

Researchers from Belgium reported that filamentous fungi are able to produce fruity or floral odors: "White rot fungi are known to metabolize ferulic acid into vanillic acid and vanillin... the fungal plant pathogen *Botryodiplodia theobromae* can also form methyl(+)-7-iso-jasmonic acid, which displays a sweet floral, jasmine-like odour."[2] The study also notes that rose and even galbanum scents can be created using yeast fermentation by *Saccharomyces cerevisiae* and *Kluyveromyces marxianus*.

Perhaps the worst news about these cutting-edge technologies is that according to FDA rules and European regulations, since the flavor molecule comes from an edible yeast, it is considered "natural."

Seed to Seal is your guarantee that Young Living labs will reject fake vanilla-burping yeast additions that attempt bio-adulteration. Young Living provides essential oils that are the true essence of the plant or tree and are completely pure.

1. Carroll AL, et al. Microbial production of scent and flavor compounds. *Curr Opin Biotechnol*. 2015 Sep 28;37:8-15.
2. Vandamme EJ. Bioflavors and fragrances via fungi and their enzymes. *Int J All Facets Mycol*. 2003 13:415-421.

IRMS (isotope ratio mass spectrometer) detects synthetic chemicals to ensure Young Living's essential oil quality.

Fume hood for conducting flash point testing.

UHPLC (ultra high pressure liquid chromatography) tests vitamins and components of dietary supplements and foods.

Refractometer for measuring refractive index.

Checking for bacteria.

Conducting micro testing in the Quality Control laboratory.

Loading oils in the Autosampler of the GC/MS instrument.

Analyzing the GC/MS report.

Young Living Global Headquarters in Lehi, Utah, since 2003.

YOUNG LIVING TODAY

As of July 2015, Young Living reached a billion dollars in sales with over one million members in 133 countries throughout the world. The global headquarters are in Lehi, Utah, with manufacturing in Spanish Fork, Utah. The international offices in Canada, Australia, Singapore, Japan, Malaysia, United Kingdom, Ecuador, Mexico, Croatia, and the United States currently employ about 2,500 people.

Young Living has eight farms and distilleries in operation in six countries and many partner and cooperative farms worldwide that supply the essential oils that are the foundation of where our Seed to Seal process begins.

The Young Living warehouse in Utah has a state-of-the-art, pick-to-light order-filling system that ships between 18,000 and 24,000 orders daily. Currently, the manufacturing department has a filling capacity of 300,000 bottles of oil a day, but that will soon increase dramatically with the new warehouse expansion that is expected to be in operation by March 2016. The new expansion will enable manufacturing to reach a capacity of 500,000 bottles daily.

This is a most remarkable legacy for the man who came from such simple beginnings. Wanting to have something better than what his parents had and to live in a better environment than that in which he grew up, Gary Young is an amazing example of perseverance and determination. He believes that all things are possible, and those desiring and willing to pay the price can be successful and fulfill their dreams.

"No other person in the world has gone into the unknown reaches of the world, sometimes at great risk, in so many countries and continents, and with such a wide variety of climates, looking for plants to grow, harvest, and distill than D. Gary Young," says Jean-Noël Landel.

Young Living's manufacturing and bottling plant, laboratory testing, quality control, state-of-the-art, pick-by-light packing and shipping since 2007. It is currently expanding to 200,000 square feet, projected to be completed early 2016.

Young Living International Grand Convention, 2015.

Australia headquarters, established in Brisbane, February 1999.
The new Sydney office opens January 2016.

Small meeting room in the new Sydney office.

Thousands of members were excited to attend the International Grand Convention in Salt Lake City in 2014.

Japan headquarters, 32nd floor, Tokyo, March 2011. The entrance is covered with breathtaking dried lavender flowers that were flown in from the Young Living farm in France.

Malaysia headquarters, Kuala Lampur, October 2014.

Singapore office, October 2011.

Ecuador headquarters, Guayaquil, November 2006.

Entrance and reception area in the Ecuador Office.

European headquarters, established April 2005. The new London office opens February 2016.

Canada headquarters, Calgary, March 2013.

Mexico headquarters, Mexico City, August 2009.

Hong Kong, established October 2013. The new office is on the 6th and 7th floors of this beautiful building that opened in November 2015.

Gary's heart is filled with tremendous satisfaction as he watches the first drops of Black Spruce oil bubble up in the separator during the first distillation at the Northern Lights Farm in Canada, March 2014.

A MODERN-DAY PIONEER

Gary Young is a truth seeker, and he follows the path that will lead him there. He is not one to write about someone else's adventures. He must know and see for himself. He writes about his own experiences, research, and findings.

His determination has taken him to Shabwa, the ancient dwelling of Queen Sheba, hidden from the world in the "Forbidden Zone" of Yemen; to the unknown tribal land of frereana frankincense in Somalia; to the tops of the Al-Hasik Mountains in Oman; deep into the jungles of the Amazon; to Sri Lanka to find the elusive blue lotus; and to many other previously unexplored places looking for the hidden secrets of God's oils.

As Gary has said, his mistakes are numerous, but they have all taught him more about life and what works and doesn't work. He has been misled and deceived in his desire to help others, which has taught him greater understanding and perhaps tolerance. He has lived on the edge of life with pain and suffering that has taught him endurance and persistence. He has ventured into unknown danger seeking truth, giving him courage and gratitude for God's blessings in his life. His adventures are an open book, and he eagerly shares his discoveries with those who will listen and benefit from his knowledge and experience. We are grateful that his life's story has been preserved in the photos of this book.

He grew up farming and living off of what Mother Nature offered, which has given him a tremendous advantage in his life's path. He has invested 34 years in the research and discovery of essential oils and their benefits and 29 years learning how to blend oils for enhanced benefits and infusing them into food supplementation.

In addition, he has spent over 25 years farming aromatic plants, learning distillation and analytical evaluation, and traveling to the far corners of the earth in his study of essential oils. He has designed and built 8 large distillation facilities around the world, with others in the planning stages. No other company has the history of research and development of essential oils in the areas of production, usage, research, and education as D. Gary Young and his company, Young Living Essential Oils.

Members of Young Living, the World Leader in Essential Oils, spread the knowledge of God's precious oils to every corner of the globe, giving hope to the needy and an opportunity for those looking for a way to help our world physically, spiritually, and financially.

For more than 20 years, thousands of Young Living members have come to one or more of the farms to help with a harvest and to expand their understanding and their appreciation as they learn what it takes to produce an essential oil. The camaraderie, sharing, working together, and helping each other in this kind of work in these conditions cannot be experienced anywhere else in the world. It is truly a life-changing experience.

Young Living sets an example in the network marketing industry in the integrity of its products and a compensation plan that makes it possible for those interested to realize their dreams. It creates an honest environment where its members are invited to be a part of the Seed to Seal process, so they can speak with a knowing of the value of their products and from where they come.

D. Gary Young is truly a pioneer in essential oil research, an inventor of equipment, a developer of the process, a scientist and formulator in the laboratory, an educator on the cutting edge of discovery, and a leader in usage and application, who has taught millions of people in many countries.

He is a writer with deep human insight and understanding, a man with an inner knowing of God's great gift of essential oils to mankind.

Today, with the education and experience of almost 30 years in growing crops for essential oil production and designing, building, and operating distilleries around the world, D. Gary Young has certainly earned the distinction of "The World Leader in Essential Oils."

The Many Facets of D. Gary Young

The life of D. Gary Young has many facets. He is a man who is comfortable wherever he is and whatever he is doing, from riding his horse in the mountains, to formulating in the lab, to teaching thousands of people from stage. His life is an inspiring story of what most people would say is impossible—to overcome monumental obstacles and succeed. We hope this book about his life's journey has warmed your heart, motivated you, and excited you to "go for it" and live your dream.

1st Place, Over 50, 2002.

Publications by D. Gary Young, 1996-2015

Co-Authored Research

Detecting Essential Oil Adulteration. Boren KE, Young DG, Woolley CL, Smith BL, Carlson RE. J Environmental Analytical Chemistry. 2015 2:2.

Differential effects of selective frankincense (Ru Xiang) essential oil versus non-selective sandalwood (Tan Xiang) essential oil on cultured bladder cancer cells: a microarray and bioinformatics study. Dozmorov MG, Yang Q, Wu W, Wren J, Suhail MM, Woolley CL, Young DG, Fung KM, Lin HK. Chinese Medicine. 2014 Jul 2,9:18.

Management of basal cell carcinoma of the skin using frankincense (Boswellia sacra) essential oil: a case report. Fung KM, Suhail MM, McClendon B, Woolley CL, Young DG, Lin HK. OA [Open Access] Alternative Medicine. June 01, 2013 1(2):14.

Extraction of biologically active compounds by hydrodistillation of Boswellia species gum resins for anti-cancer therapy. Lin HK, Suhail MM, Fung KM, Woolley CL, Young DG. OA (Open Access) Alternative Medicine, 2013 Feb 02;1(1):4.

Chemical differentiation of Boswellia sacra and Boswellia carterii essential oil by gas chromatography and chiral gas chromatography-mass spectrometry. Woolley CL, Suhail MM, Smith BL, Boren KE, Taylor LC, Schreuder MF, Chai JK, Casabianca H, Haq S, Lin HK, Al-Shari AA, Al-Hatmi S, Young DG. Journal of Chromatography A. Oct 2012.

Frankincense essential oil prepared from hydrodistillation of Boswellia sacra gum resins induces human pancreatic cancer cell death in cultures and in a xenograft murine model. Ni X, Suhail MM, Yang Q, Cao A, Fung K-M, Postier RG, Woolley CL, Young DG, Zhang J, Lin HK. BMC Complementary and Alternative Medicine. Dec 2012.

Boswellia sacra essential oil induces tumor cell-specific apoptosis and suppresses tumor aggressiveness in cultured human breast cancer cells. Suhail MM, Wu W, Cao A, Mondalek FG, Fung K-M, Shih P-T, Fang Y-T, Woolley CL, Young DG, Lin HK. BMC Complementary and Alternative Medicine. 2011.

Inhibition of methicillin-resistant Staphylococcus aureus (MRSA) by essential oils. Chao S, Young DG, Oberg C, Nakaoka K. Flavor and Fragrance Journal. Vol. 23 2008.

Essential Oil of Bursera graveolens (Kunth) Triana et Planch from Ecuador. Young DG, Chao S, Casabianca H, Bertrand M-C, Minga D. Journal of Essential Oil Research. Nov/Dec 2007.

Inhibition of LPS Induced Nitric Oxide Production in Murine RAW Macrophage-like Cells by Essential Oils of Plants. Chao S, Young DG, Nakaoka K, Oberg C. Journal of the Utah Academy of Sciences, Arts, and Letters. Vol. 82, 2005.

Assessment of Antimicrobial Activity of Fourteen Essential Oils When Using Dilution and Diffusion Methods. Donaldson JR, Warner SL, Cates RG, Young DG. Pharmaceutical Biology. Vol. 43, No. 8. 2005.

Pre-Clinical Study: Antioxidant Levels and Immunomodulatory Effects of Wolfberry Juice and other Juice Mixtures in Mice. Chao S, Schreuder M, Young DG, Nakaoka K, Moyes L, Oberg C. Journal of the American Nutraceutical Association. Winter 2004.

Composition of the Oils of Three Chrysothamnus nauseousus Varieties. Chao S, Young DG, Casabianca H, Bertrand M-C. Journal of Essential Oil Research. Nov/Dec 2003.

Antimicrobial Effects of Essential Oils on Streptococcus pneumoniae. Horne D, Holm M, Oberg C, Chao S, Young DG. Journal of Essential Oil Research. Sep/Oct 2001.

Screening for Inhibitory Activity of Essential Oils on Selected Bacteria, Fungi and Viruses. Chao SC, Young DG, Oberg CJ. Journal of Essential Oil Research. Sep/Oct 2000.

Effect of a Diffused Essential Oil Blend on Bacterial Bioaerosols, Chao SC, Young DG, Oberg CJ. Journal of Essential Oil Research. Sep/Oct 1998.

Papers Presented at Conferences

Inhibitory Activity of Essential Oils Toward Cellular Proliferation in a Lung Cancer Carcinoma Cell Line. International Food Chemistry Conference. SICC-5, Singapore, 2007.

Comparative Study of the Essential Oils of Three Betula Species: B. alleghaniensis, B. lenta, B. pendula. Gualin, China, Conference, 2004.

Autolysis Induction by Essential Oils in Streptococcus pneumoniae. Annual Meeting of the Utah Academy of Sciences, Arts, and Letters. Salt Lake City, UT, April 1998.

Cultivating and Distilling Therapeutic Quality Essential Oils in the United States. Proceedings from the First International Symposium, Grasse, France. March 21-22, 1998.

Organic Farming and Germination: UNIDO World Congress on Essential Oils, Anadolu University in Eskisehir, Turkey, 1997.

Books Published

Ancient Einkorn: Today's Staff of Life – 2014
Shutran's Ancient Apothecary – 2011
The One Gift – 2010
Raindrop Technique – 2008
Discovery of the Ultimate Superfood – 2005
Essential Oil Integrative Medical Guide – 2003
A New Route to Robust Health – 2000
Pregnenolone – 2000
Longevity Secrets – 1999
The Truth Behind Growth Hormone – 1999
An Introduction to Young Living Essential Oils – 1999
Aromatherapy: The Essential Beginning – 1995